U.S. ENERGY POLICY

U.S. ENERGY POLICY

CRISIS AND COMPLACENCY

By DON E. KASH
and ROBERT W. RYCROFT

UNIVERSITY OF OKLAHOMA PRESS : NORMAN

BY DON E. KASH

(with others) *Energy Under the Oceans: A Technology Assessment of Outer Continental Shelf Oil and Gas Operations* (Norman, 1973)

(with Robert W. Rycroft and others) *Our Energy Future: The Role of Research, Development, and Demonstration in Reaching a National Consensus on Energy Supply* (Norman, 1976)

(with Robert W. Rycroft) *U.S. Energy Policy: Crisis and Complacency* (Norman, 1984)

BY ROBERT W. RYCROFT

(with Don E. Kash and others) *Our Energy Future: The Role of Research, Development, and Demonstration in Reaching a National Consensus on Energy Supply* (Norman, 1976)

(with others) *Energy Policy-Making: A Selected Bibliography* (Norman, 1977)

(with Joseph S. Szyliowicz) *Decision-Making in a Technological Environment: The Case of the Aswan High Dam* (Boston, 1980)

(with James E. Monaghan) *Cumulative Impacts of Energy and Defense Projects in the West: Synfuels and the MX* (Columbus, Ohio, 1981)

(with Robert D. Brenner) *Nuclear Energy Facility Siting Policy in the United States: Implications of the International Experience* (Princeton, N.J., 1981)

(with others) *Energy from the West: A Technology Assessment of Western Energy Resource Development* (Norman, 1981)

(with others, coeditor) *Energy and the Western United States: Politics and Development* (New York, 1982)

(with Don E. Kash) *U.S. Energy Policy: Crisis and Complacency* (Norman, 1984)

Library of Congress Cataloging in Publication Data

Kash, Don E.
  U.S. energy policy.

  Includes bibliographies and index.
  1. Energy policy—United States.   2. Energy policy—United States—History.
I. Rycroft, Robert W.   II. Title.   III. Title: US energy policy.
HD9502.U52K375   1984   333.79'0973   83-17093
(alk. paper)

The paper in this book meets the guidelines for permanence and durability of the Committee on Production Guidelines for Book Longevity of the Council on Library Resources, Inc.

To Vernon Van Dyke
   For those of us who were his students
   the standard that he set has been the goal.

Contents

CONTENTS

Illustrations

Preface

The period since the oil embargo of 1973 has seen a continuing stream of analyses of the nation's energy situation and prognostications about its energy future. We, in fact, have contributed to that stream with a number of books and articles. From hindsight, what has been most striking about the continuing body of analysis is the consistency with which it has been proved incorrect by later events.

We undertook this book with the record of inadequacy in energy policy analysis uppermost in our minds. Our purpose is to analyze the nation's recent energy history and specifically the actions taken to formulate a national energy policy in response to the energy crisis. Our starting assumption was that the United States policy process had failed to respond effectively to that crisis.

In the course of doing the research for this book, we found that our initial perception of policy failure was incorrect. Although the process of evolving an energy policy was fragmented, chaotic, and incremental, by the end of 1980 the United States had a workable energy policy. We found that between 1973 and 1980 the United States policy system had been developing a new national consensus on energy. That consensus existed, at least in rudimentary form, when the Reagan administration came into office.

What occurred between the oil embargo and the advent of the Reagan administration was a process of new policy formulation similar to what had occurred whenever the nation had faced policy crises in the past. In the American polity, successful national policy requires the development of a national consensus. Because of its importance and complexity, and because the energy crisis was a historical turning point for Americans, consensus was slow in coming. In truth, most Americans fail to recognize the distinctive accomplishments of the period from 1973 to 1980. As we have noted, we also failed to recognize it.

This book thus represents a revision of the conventional

wisdom held by most laymen and most energy specialists about what had occurred with regard to energy policy by 1980. The failure to understand both how the national policymaking process works and what it had accomplished before the advent of the Reagan administration has and will have costly consequences for our society.

As the book explicates in some detail, those costs are clearly represented in the radical rejection by the Reagan administration of the nation's achievement in energy policy. In moving the management of energy into the marketplace, the Reagan administration has rejected the fundamental tenet of the conservative: it has rejected both the procedures and the mechanisms that for over two hundred years have provided national stability.

This book concludes with a set of recommendations for actions that we believe are necessary if the nation is to have a stable energy future. We believe that energy is so central to the nation's welfare that it should not be used as an instrument for experimenting with untried concepts.

Don E. Kash
Robert W. Rycroft

Norman, Oklahoma

Acknowledgments

This book could not have been completed without the assistance and support of many people. In particular our colleagues in the Science and Public Policy Program of the University of Oklahoma provided a fertile environment for the discussion of energy policy developments. Joye R. Swain typed and critiqued an untold number of versions and was invaluable. Sandy Seay maintained incredibly good humor while digging through masses of documents. The following members of the staffs of the Science and Public Policy Program of the University of Oklahoma and the Graduate Program in Science, Technology, and Public Policy in George Washington University deserve special thanks. They are Mary Zimbleman, Ellen Ladd, Lennet Bledsoe, Kimberly Lutz, Lucienne Beard, and Ginger Keller.

Thomas J. Willbanks, Richard Rowberg, Larry Parker, Martin Cines, Harry Perry, Steve Ballard, Mike Devine, Betsy Gunn, and Vernon Van Dyke read and critiqued all or portions of the manuscript. Good criticism is an essential ingredient in any research effort, and we benefited greatly from the time these individuals devoted to the manuscript. The reviewers obviously have no responsibility for factual errors, conclusions, or recommendations.

We also thank Ron Burton and the University of Oklahoma Foundation and Kenneth L. Hoving, Vice-Provost for Research in the University of Oklahoma, for their support.

Finally, special thanks go to Bev and Marilyn for their support and tolerance.

Don E. Kash
Robert W. Rycroft

Norman, Oklahoma

Abbreviations Used in This Book

AEC	Atomic Energy Commission
AGA	American Gas Association
API	American Petroleum Institute
Aramco	Arabian American Oil Company
Btu	British thermal unit
DENR	Department of Energy and Natural Resources
DOE	Department of Energy
DOI	Department of the Interior
EEI	Edison Electric Institute
EIS	Environmental-impact statement
EMB	Energy Mobilization Board
EPA	Environmental Protection Agency
EPRI	Electric Power Research Institute
ERDA	Energy Research and Development Administration
FEA	Federal Energy Administration
FLPMA	Federal Lands Policy and Management Act
FPC	Federal Power Commission
FY	Fiscal year
HTGR	High-temperature gas reactor
IUOE	International Union of Operating Engineers
LNG	Liquefied natural gas
LMFBR	Liquid-metal fast-breeder reactor
LWR	Light-water reactor
MBPD	Million barrels per day
MCF	Thousand cubic feet
MESA	Mine Enforcement and Safety Administration
MHD	Magnetohydrodynamics
NCA	National Coal Association
NEPA	National Environmental Policy Act
NPC	National Petroleum Council
NPR	Naval Petroleum Reserve
NPRA	National Petroleum Reserve Alaska
NRC	Nuclear Regulatory Commission
OCAW	Oil, Chemical, and Atomic Workers Union
OCS	Outer continental shelf

OECD	Organization for Economic Cooperation and Development
OPEC	Organization of Petroleum Exporting Countries
ORNL	Oak Ridge National Laboratory
OSM	Office of Surface Mining
PIES	Project Independence Evaluation System
Pu-239	Plutonium-239
R&D	Research and development
SERI	Solar Energy Research Institute
TAPS	Trans-Alaska Pipeline System
TCF	Trillion cubic feet
Th-232	Thorium-232
TVA	Tennessee Valley Authority
USGS	U.S. Geological Survey
U-233	Uranium-233
U-235	Uranium-235
U-238	Uranium-238
U_3O_8	Yellowcake
UF_6	Uranium hexaflouride
UMW	United Mine Workers
USGS	United States Geological Survey

PART ONE

Traditional Energy Policy Formulation

CHAPTER 1

An Overview of Energy Policy

This [energy] perhaps has been the most parochial issue that could ever hit the floor. —Speaker of the House of Representatives Thomas P. O'Neill, Jr., 1975

Our decision about energy will test the character of the American people and the ability of the President and the Congress to govern. This difficult effort will be the "moral equivalent of war." —President Jimmy Carter, April 18, 1977

Urgent calls for policy action on energy and expressions of frustration over inaction characterized the 1970s. From 1973 through 1980 three presidents and four Congresses struggled continuously to formulate an effective energy policy for the nation. Energy commanded a front and center position on the domestic and international political stage (Schurr et al., 1979, p. xxi).

The period since 1980, however, has been quite a different story. President Ronald Reagan rejected the view that energy deserved a high place on the policy agenda. Thus, in the words of David Davis (1982, p. 2), the Reagan administration has "sought to dismantle, decontrol, and deregulate organizations, policies, and regulations put in place during the 1970s." Indicators of this change in priorities abound, but perhaps the most dramatic evidence lies in the very different patterns of federal support for programs of the Department of Energy (DOE). As figure 1.1 indicates, the Reagan administration's funding for fiscal year (FY) 1982 and its budget proposals for FY 1983 represented a marked departure from previous policy.

How, in the course of a decade, could a public policy issue be seen as the "most parochial issue to hit the floor," the "moral equivalent of war," and a low-priority concern? The answer lies in the pervasive importance of energy, on the one hand, and in its equally pervasive complexity, on the other.

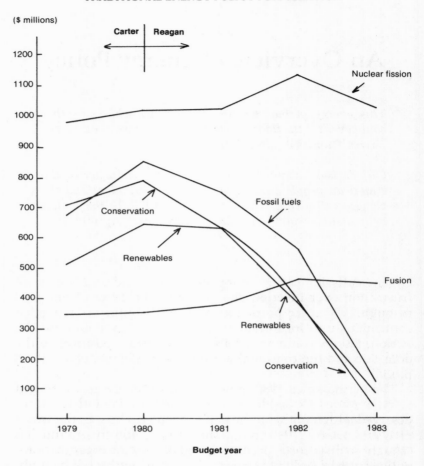

Fig. 1.1. Trends in federal energy funding. From Friends of the Earth, 1982, p. 43.

The political salience of energy is obvious. Cheap, abundant energy has been the very life blood of America's industrial-technological society. The energy shortfall of the early 1970s and the rapid escalation in price immediately affected nearly every American. To the extent that they were able, Americans put pressure on the president and the Congress to protect their special interests. Taken together, these diverse and conflicting demands on government precluded the rapid formulation of a

4

national energy policy. The pressure for a national energy policy was unremitting. Yet there was nothing resembling a national consensus on what that policy should be or how it should be formulated and implemented. The problem was not an absence of policy choices but rather the surfeit of energy alternatives. One study characterized the energy situation as follows: "The combination of possible decisions is enormously complex and the menu before us is replete with dilemmas" (Schurr et al., 1979, p. 2).

This book investigates why events in the early 1970s and especially the 1973 oil embargo triggered an energy crisis in the United States, how the nation dealt with that crisis in the decade of the 1970s, how the Reagan administration sought to reverse the nation's approach to energy policymaking and implementation, and what must be done to resolve energy problems and issues in the years to come.

The major cause of the energy crisis was the inappropriate fit between the facts of energy and policy for energy. Energy is "a scientific concept by which the basic processes of nature, whether physical, chemical, or biological, are rationalized" (Crabbe and McBride, 1979, p. 62). Energy is the "capacity for doing work," which has come to be measured in terms of British thermal units (Btu's). Policy refers to "efforts in and through government to resolve public problems" (Jones, 1977, p. 5). Energy policy, then, refers to those actions taken by and through government either to produce a supply of or to reduce the demand for Btu's.

By the early 1970s the nation's policies for energy did not match the physical circumstances of the nation's energy system. In the words of Barnett (1980, p. 6), there was a "misfit between politics and the natural order which neither economists nor corporate executives nor government bureaucrats quite understand." This situation had been building for some time, but it took the shock of the oil embargo of 1973 to bring home the degree to which policy had become divorced from reality.

In the years preceding the embargo, energy policy processes in the United States were organized around five distinctive fuel sources—oil, natural gas, coal, nuclear power, and electricity—and the primary purpose of these policies was to manage surplus. A fairly autonomous policy system had developed for each of these five fuels by the 1960s, and the nation's energy policy resulted from a disaggregated collection of oil, natural gas, coal, nuclear power, and electricity policy decision making. The United States did not have anything that could be termed an overall energy policy system. In other words:

Until the 1973 Arab Embargo, the five resource subsystems were relatively self-contained decision making communities, each with a fairly stable set of participants and decision-making procedures. This is not to say that decisions in these subsystems were at all centralized or characterized by comprehensive planning. Quite the contrary, each resource development subsystem had its own unique form of problem solving which permitted it to cope with situations in which goals, alternatives, consequences and even the problems themselves were often undefined. (Kash et al., 1976, p. 48)

By 1973 the country was importing more than 30 percent of its oil, and a portion of that oil came from "insecure" suppliers. Yet no policy existed for dealing with a situation in which imports might be denied. The fuel systems had developed when there was a domestic surplus of all five fuels, and they were unable to respond to the radically different energy environment of the 1970s.

The embargo demonstrated the inadequacy of the fuel policy systems in a dramatic fashion. The nation found itself unable to respond comprehensively and systematically to the denial of petroleum; thus the energy crisis was triggered. Immediately there was a widespread perception of the need to integrate the fuel systems, but no policy capability existed to bring about this integration. For the first time in its history the United States was forced to think in terms of fuel trade-offs, and energy was placed on the national policy agenda. The president and the Congress were faced with such concerns as finding substitutes for oil; linking and coordinating highly sophisticated energy research, exploration, production, transportation, and utilization activities; and examining a spectrum of the costs, risks, and benefits of a wide range of energy choices.

In the immediate (1973–75) postembargo period the need for a comprehensive approach to energy received more symbolic than substantive attention. As with any other new item on the national policy agenda, there was lag time between the perception of the problem and the ability to define it in meaningful policy terms. In the interim it was important for govern-

6

ment to be able to demonstrate that it was "doing something," even if in only a symbolic way. The period was characterized by self-sufficiency rhetoric such as "Project Independence" and calls for scientific crash programs and technological fixes, such as synthetic-fuels projects, but very little action. Underlying much of the rhetoric was a perception that the energy crisis was a relatively short-term difficulty, amenable to an engineering solution. This optimism began to erode as a broader understanding of energy and its complexities developed. By 1975 the talk of self-sufficiency had lessened dramatically. Public response to such programs as Project Independence had been apathetic at best, and rhetoric had failed to achieve the dual policy objectives of reducing United States oil imports and increasing domestic production of all forms of energy.

Once the period of initial symbolic response had passed, the executive and legislative branches of the federal government began to address energy problems in a more substantive way. The president and the Congress quickly discovered, however, that there was little agreement on even the most fundamental facts of energy (Schurr et al., 1979, pp. 4–12). Thomas Wilbanks (1981, pp. 2–3) has sketched the nation's dilemma as follows:

This, then, is our energy problem: We are vulnerable but divided, when reducing our vulnerability requires that we be able to act with a kind of purpose and determination that, in our system, requires a very broad consensus. We need to begin a long and difficult transition, but we are not able to agree to what, in which directions, and with what first steps. Clearly our alternatives are many, the uncertainties are great, and the participants in energy policy making have a wide range of legitimate concerns. But if we fail to identify a path that we can travel together, the result will almost certainly be disastrous. The American people expect our economy and our system of government to deliver the energy it takes to maintain a good standard of living. If we fail to deliver, there are likely to be significant and irreversible changes in our system.

To appreciate this dilemma, it is necessary to recall the assumptions that most Americans had about energy and the way in which those assumptions were assaulted in 1973 and the years that followed. Abundant energy delivered at declining prices was a national expectation, viewed almost as an American birthright. This assumption was not derived from some abstract philosophy. Rather, as shown in table 1.1, it was a consequence of the American experience. For more than one hundred years the nation's use of energy had doubled or nearly doubled every twenty to twenty-five years. As energy use increased, the price of energy declined, in some forms at an astonishing pace. For example, in 1973 residential electricity consumers were paying only 20 percent of what they had paid in 1940 for the same amount of electricity (Ross and Williams, 1981, p. 9).

Table 1.1. U.S. Energy Consumption, 1850–1976

Year	Total Energy Consumed (10^{12} Btu)	Total Energy per Capita (Million Btu)
1850	2,500	105
1900	8,300	110
1920	19,782	186
1925	20,809	180
1930	22,288	181
1935	19,107	150
1940	23,908	181
1945	31,541	238
1950	34,153	226
1955	39,956	243
1960	44,816	249
1965	53,969	278
1970	67,444	330
1972	72,108	345
1973	75,561	359
1974	73,941	346
1976	74,500	342

Source: Dorf, 1978, p. 2.

The luxury of cheap, abundant energy allowed the producers and regulators of the five fuels to develop their own distinct policy frameworks independent of more general energy or societal concerns. The differences among the systems reflected the varying physical characteristics and historical development of the fuels. Given exponential growth in demand, the interaction among the systems was not a serious problem. Marketplace competition during a period of rapid growth in demand and surplus production capability allowed each fuel system a high degree of autonomy and provided a smooth working environment among the systems. In such a context the formulation of policy for and the management of each fuel could be left to a relatively narrow group of actors who had a vested interest in the production and delivery of the fuel. As long as the fuels were produced in abundance at declining costs, these policy participants remained essentially free from outside interference.

By the late 1960s and early 1970s the nation's energy situation was undergoing rapid and massive change. Energy consumption had grown at a rate of 3.5 percent a year for fifteen years after 1950, and that rate increased to 4.5 percent between 1965 and 1973. Domestic production of low-cost energy, however, was not able to keep pace with this growth in demand. The gap between consumption and domestic supply increasingly was being filled by low-cost imported petroleum. Unfortunately for the continuation of this trend, by 1973 world demand for oil was nearly equal to world production. In that year the United States was importing about 15 percent of its energy in the form of oil, and the percentage figure for some European countries and Japan was as high as 90 percent.

Moreover, petroleum demand in the developing world also was expanding rapidly. The gap between the supply and demand of that most preferred fuel, oil, was closing. Americans faced a situation in which both domestic surplus of all forms of energy and world capacity to produce oil over and above global demand had shrunk to almost nothing. In these changed physical circumstances existing United States policy structures were hopelessly outdated. It was this set of facts about energy, and particularly about petroleum, that made the embargo such a disruptive event.

Changes in the international energy system and the implications of those changes for domestic policy were not widely known or understood in this country before 1973. The literature on world energy production and consumption was growing, but few of these analyses had been linked to the policy process in an effective fashion. When the embargo took place, much more attention was paid to evaluations of the changing energy outlook. The picture was not pleasant. First, the United States was found to be dependent on foreign sources for 6 million barrels of oil per day (MBPD; U.S. Federal Energy Administration, 1974). Most estimates indicated that unless some kind of concerted action was taken import dependence was likely to grow. Several forecasts suggested that the United States would be importing over 10 MBPD by 1980 if these trends continued (National Petroleum Council, 1971).

As if the petroleum situation were not enough cause for alarm, analyses of natural-gas supply and demand indicated that it was only a matter of time before the country encountered severe problems with that resource as well. For several years the U.S. Federal Power Commission (FPC), which had responsibility for regulating interstate gas supplies, had been warning of an impending shortage. In 1971, for the first time, some interstate pipeline companies were unable to maintain service to all customers with whom they had firm contracts, and the winter of 1972-73 saw shortfalls during the peak heating season (U.S. House Committee on Science and Astronautics, Subcommittee on Energy, 1974). These danger signals prompted shifts to oil and electricity by some consumers, adding to the demands for those sources and demonstrating for the first time the problems that planners and decision makers would encounter in coming to grips with the complexities of interdependence among the fuel systems in a context of scarcity.

The first halting attempts at energy-system, rather than fuel-system, analysis added to the growing uncertainty about the ability of nuclear power to come to the rescue of declining fossil fuels. Recognition grew that it was not possible to substitute electricity for oil and gas without major changes in equipment. Even had the switch to electricity been possible, observers cited the lengthening lead times in constructing nuclear electric power plants. As the length of time needed to bring these complex facilities on line grew (by the early 1970s it was taking about ten years to do so), their capital costs skyrocketed. Largely owing to concerns about reactor safety and waste disposal, nuclear energy no longer appeared to be certain to replace oil and natural gas as a source of cheap, abundant energy for the future (Mancke, 1974, pp. 134-38).

The energy crisis also dramatized the critical relationship between energy and the environment. After 1973 it was no longer possible to separate energy and environmental concerns. The environment was seen as "both the source of raw materials needed to generate energy and as the depository of the pollu-

tion stemming from the production and use of energy" (Wolozin, 1974, p. v). This awareness was significant for a number
of reasons, but in the short term its prime consequence was
to place a heavy constraint on the development of United States
coal resources. For, with the more pessimistic view of the future
of nuclear power, coal was the only readily available source of
energy that could be produced rapidly in the amounts necessary to compensate for the anticipated shortfall in the other
conventional fuels. Rapidly expanding support for the goal of
a clean environment, however, hampered quick expansion of
coal-mining and combustion activities. Federal rules and regulations were implemented after the passage in 1969 of the National Environmental Policy Act (NEPA) that constrained both
the production and the use of coal in the interest of protecting
environmental quality. In the words of the preliminary report
of the Energy Policy Project of the Ford Foundation (1974a, p.
25): "From an environmental perspective, there is no entirely
satisfactory way to use coal in the near term."

The overall picture that emerged from the initial investigations of the four primary sources of energy was, in a word,
disturbing. The issues were compounded by uncertainties that
began to cloud the field of electricity generation. In fact, electric
power became one of the most widely and intensely scrutinized
elements of the nation's energy system. Especially crucial to
the assessment of the future of electricity were the very high
energy-loss characteristics associated with steam electric power
plants. For every 100 units of primary energy that go into a
steam electric facility, only about 35 units of electrical energy
come out the other end. When it was recognized, in a policy
sense, that not much could be done to change the physical
laws governing that energy loss, investigators began to question whether the use of such high-quality fuels as oil and
natural gas to generate electricity was in the national interest
(Chapman, 1974). It was noted that in the southwestern United
States a very large portion of electric power was generated with
natural gas as the fuel, while in New England scarce petroleum
served this purpose. Americans became aware that if one used
natural gas to fire a home furnace the residence received the
benefit of between 80 and 90 percent of the gas energy delivered into the furnace. Alternatively, if the home was heated
with electricity produced in a gas-fired electric power plant,

more than twice as much natural gas was required to serve the same purpose.

Something had clearly gone wrong with United States energy policies, but there was no consensus on the causes of their failure. Among the "explanations" that were offered some analysts stressed the poor choices of policy tools that had been used in the past, such as imposing regulations instead of allowing the free market to determine energy alternatives. Others argued that the failure resulted from the profligate energy-consumption patterns of Americans. And some observers saw the problem as resulting from a conspiracy to increase oil-industry profits. In retrospect it seems clear that the most significant factors were inappropriate management arrangements that had led to the misuse of high-quality sources of energy and that, plus a failure to look at energy demand, had precluded the anticipation of shortages.

The nation's institutions and processes had been far too reactive and had not developed the anticipatory capability necessary to compensate for the changing energy situation. Such a conclusion is reinforced by the recognition that United States "energy policy" in fact was merely the sum of fuel policies. No permanent mechanisms existed for monitoring and evaluating energy developments in a comprehensive manner, nor was there an organized way to assess the trade-offs among fuel policies or to resolve conflicts among the fuel systems. To cite only the most obvious example, the mismanagement of pricing policy for natural gas was a major consequence of organizing decision-making responsibility around specific fuels rather than around energy.

The natural-gas policy system had developed pricing arrangements that kept interstate gas on a Btu basis much lower than that of petroleum. This pricing policy both encouraged wasteful use of gas and discouraged exploration for and production of new gas (Roberts, 1973). This was true even though policies for oil made domestic production more costly than imported oil, while petroleum at the retail level was cheaper in the United States than throughout the rest of the world. Comparisons with European nations that had used low-cost foreign petroleum but had driven their consumers to greater conservation by levying heavy taxes on imports gave even greater emphasis to the failure of United States policies.

12

In the period following the embargo, the consciousness of these failures of policy led to calls for dramatic action to modify the nation's energy-management system. The need was to manage energy, not to control five distinct fuels. The embargo and its aftermath, therefore, shattered the United States energy framework. Stability in this framework had been based on an assumption of continuous competition among the five systems for an ever-growing market supplied by surplus fuels that could be delivered at ever-cheaper prices. Coming to grips with needed policy-system changes proved to be a difficult task, however. While there was substantial agreement that the nation must make modifications in the structural and procedural arrangements for formulating policy and managing energy (and that many of these changes were long overdue), there was little agreement about what those modifications should be.

Before broad national policy could be defined and undertaken in our highly pluralistic political system, it was necessary to develop a new consensus on the entire range of energy policy questions. In the American polity the leadership in developing a consensus falls to the president and the Congress. The nation, though badly divided and distracted by such events as Watergate, nevertheless demanded a stable policy system capable of formulating and implementing workable energy actions. The days of the highly autonomous fuel policy systems were numbered.

To a great degree the struggles over energy policy from the embargo to the end of the Carter administration can be best understood as a search for consensus. Agreement on energy matters required evolving fundamental compromises in four broad issue areas: (1) national energy policy goals, (2) future sources of United States energy, (3) appropriate policy instruments and tools to manage energy, and (4) appropriate political, organizational, and managerial mechanisms necessary for stability. Because of their importance for later chapters, each of these issue areas merits more detailed attention at this point.

ENERGY GOALS

The first step in evolving a consensus on energy policy required the determination of what that policy was supposed to

13

achieve. The four goals that formed the continuing focus of attention during the years following the petroleum embargo were abundance, cheapness, cleanness, and security. The priority ranking and precise meaning of each of these goals, however, were worked out only after long and acrimonious struggle.

Abundance

As previously noted, for decades the United States had enjoyed a surplus of energy from all domestic fuel sources. Surplus energy meant that energy producers used every opportunity to encourage its use to replace more expensive or inconvenient ways of accomplishing tasks. In a time of abundance energy supply pushed energy demand. The changing facts of energy, spotlighted by the embargo, made it clear that in the future surplus energy would be available only as a result of conscious national policy. Otherwise the nation was faced with having to redefine what "abundant energy" meant.

Cheapness

Low-cost energy, like abundance, also had been a fact of life. The availability of ever-larger quantities of energy at decreasing prices allowed the nation to ignore two criteria that had been fundamental in less fortunate societies: efficiency and equity. As long as the energy pie was expanding and the unit price of energy was declining, economic efficiency did not require energy efficiency. The United States could afford to use energy in wasteful, inefficient ways. Similarly, equity was not an issue when the price of energy dropped each year. The poor as well as the rich benefited from price declines.

As with abundance, the facts that came to light following the embargo suggested that if the nation was to have cheap energy in the future it must embark on a conscious policy course. To do so would not be easy, however. Decisions about cheap energy were fraught with arguments regarding efficiency and equity in the years after 1973.

Cleanness

Unlike abundance and cheapness, clean energy sources had

14

not been a traditional concern of Americans. Throughout most of United States history a particular focus on the environmental consequences of energy production and consumption was lacking. In the late 1960s, however, after the publication of several landmark books, such as Rachel Carson's *Silent Spring*, and a series of widely publicized energy-related accidents, such as the Torrey Canyon spill and the Santa Barbara blowout, environmental protection became a national goal. Incorporating environmental protection into energy decision making produced a number of painful and controversial struggles. By the early 1970s environmental protection had become a national goal backed by major legislation, such as NEPA and the Federal Water Pollution Control Act. Support for environmental protection became such a powerful force that in the years following the embargo the debate over energy policy often was framed in terms of an energy-environment conflict. Tension between energy and environment was focused on the so-called bridging fuels, coal and nuclear power, which were perceived by many as being the most threatening to the environment. Thus the most obvious solutions to the oil shortage were least acceptable environmentally.

Security

The oil embargo made security a primary concern of energy policy. Like the first three goals that competed for high priority in the formulation of energy actions, "security" had various meanings for spokesmen. For example, some called for total domestic self-sufficiency, while others wanted to leave energy totally in the hands of the marketplace. The changing definitions and priority ranking of security contributed to the difficulty the nation faced in formulating an energy policy.

In sum, before a consensus could be developed for energy policy in the United States, it was necessary to give more precise meanings to the four goals, as well as to determine their priority rankings.

ENERGY SOURCES

The easiest and most attractive answer to either real or perceived shortages of energy (a distinction that has always been a major burden for energy policymakers) was to find new sources. One response to the national energy crisis had an oft-repeated refrain: "If we can put a man on the moon, why can't we . . . ?" In fact, the immediate reaction of the president and Congress was to look for new sources of energy, and in particular for energy that could be supplied by new technologies. A plethora of technological fixes was offered. Almost overnight the number of possible new energy resources doubled. Added to the list of energy resources that the nation was already using were oil shale, tar sands, geothermal resources, organic wastes, solar energy, and conservation, as well as new technologies designed to use conventional fuel sources.

Unfortunately, most of the newly proposed alternatives were characterized by a high degree of uncertainty. They were not commercially available "off the shelf." That is, there was no agreement within the scientific-technical community about when these proposed new sources could be made available, at what price, or with what environmental consequences. The wide range of options, surrounded by technical, economic, and ecological uncertainties, complicated the process of arriving at a consensus on energy policy.

POLICY INSTRUMENTS

Both the executive and the legislative branches of the federal government spent much of the postembargo period experimenting with many policy instruments and tools in an effort to develop an energy policy apparatus that would produce acceptable outcomes. These instruments ranged from exhortations by political leaders that Americans conserve energy through governmental attempts to regulate the price of various fuels to an ultimate decision to decontrol most resources. Along the way efforts were made to (1) develop performance and environmental standards, (2) employ economic sanctions to promote specific energy-use behavior, and (3) provide energy subsidies (including tax credits and direct federal outlays). Equally significant, the national government sought to underwrite a massive program in energy research and development.

POLITICAL, MANAGERIAL, AND ORGANIZATIONAL RESPONSES

As described above, before the embargo the formulation and implementation of policy in the fuel systems was the responsibility of relatively autonomous groups of actors and interests. In general, the key players were energy-producing companies and government agencies. In the early 1970s environmentalists and government organizations concerned with the environment began playing significant roles in the various fuel policy systems.

After 1973 the number of interests seeking to influence energy policy expanded exponentially. Nearly every economic interest group, ranging from farmers to truckers to homeowners, sought to ensure that their concerns were represented as the nation struggled to formulate and carry out a new energy policy.

Calls for organizational change were promulgated by this diverse set of interests as fast as they entered the struggle over energy policy, the most common demand being for centralization and policy comprehensiveness. Thus the Ford Foundation's Energy Policy Project (1974, p. 41) reflected the dominant view when it called for the designation of "a single permanently constituted body with national energy policy oversight and coordination responsibility."

One of the major difficulties in responding to organizational and managerial pressures was the need to do two seemingly contradictory things at the same time. While the concentration of responsibility for all forms of energy was an attractive idea, such a reform would make it very difficult to meet a second set of pressures: demands for broad public participation in energy matters. From 1973 through 1980 the nation experienced a major sorting-out process, determining who would participate in energy policymaking and what the organizational arrangements for citizen involvement would be. It took nearly seven years for the basic outlines of this political, managerial, and organizational framework to begin to emerge.

By 1980 the president and the Congress had been able to reach compromises on the four basic issues that faced them following the onset of the energy crisis. With decisions on these four issues the foundations for a stable national energy policy system appeared to be in place. The beginnings of an elementary consensus offered the nation the opportunity to establish for energy the same kind of policy system that existed before 1973 for each of the fuels. The rudimentary energy policy system that was in place by 1980 provided the framework necessary to manage both energy supply and demand and to develop new resources.

REJECTING TRADITION

Over the period from 1973 to 1980, then, the policy system was in the business of doing what has been done whenever the nation has faced a policy crisis. The president and the Congress, in conjunction with a range of parties-at-interest, established the basis for the stable management of important physical activities. In the years after 1980, however, the Reagan administration disrupted and eroded the fragile energy policy consensus. In a major departure from three previous presidents, two Republican and one Democratic, Reagan rejected the need for an energy policy and declared the resolution of most energy matters to be the province of the private sector. Reagan's rollback of achievements in energy policy development eliminated the nation's capacity for stable, long-term energy management.

This book argues that the critical importance of energy to the nation requires a stable energy policy system. By advocating the movement of energy management to the "free market," the Reagan ideology claims to maximize individual preferences, permit more efficient resource utilization, and maximize supply (see Palmer and Sawhill, 1982). In reality, however, this ideology violates the nation's traditional approach to the management of important physical activities and leads to a significant shift in the locus of decision making in American society. As Edgmon (1979, p. 89) has argued:

In theory, the market approach maximizes individual preferences, but such choice maximization and resource utilization efficiencies are a function of the organization of energy suppliers and the public's perception of the nature of the good being exchanged. If competitive markets existed in all fuel areas, and few opportunity costs were associated with switching from one fuel to another, then such a basis for energy organization might be viable. However, producers and distributors of energy represent highly concentrated and centralized organizations for which market principles may not strictly apply. Also, fuels and electricity may not be considered to be private goods subject to disposition and consumption through the exercise of private preferences. Society perceives values other than Btus and kilowatts associated with energy decision making. This fact is evident in the numerous social and environmental issues raised in relation to the making of energy policy decisions.

The massive shifts in power and economic resources that are being triggered by the Reagan strategy of decontrol and deregulation cannot take place without conflicts. This is particularly the case if alternative values and parties-at-interest are not incorporated into the policy process. Thus it is not surprising that the period since 1980 has featured a return to the politics of the 1973–80 period, a politics of confrontation, controversy, and declining trust in the major energy institutions and actors.

The Reagan administration effectively has returned energy to the same kind of environment that existed immediately following the embargo. That is, United States energy policymaking has been returned to the days when veto politics and reliance on the courts were dominant (Garvey, 1975).

This is not to argue that the marketplace itself leads to instability but to assert that the Reagan administration's implementation of this policy shows a lack of appreciation for the pluralistic nature of American society, a tendency to resort to black-and-white thinking about energy problems and issues, and a preference for simplistic solutions to very complex concerns. The Reagan approach fails to appreciate the need of the energy market for stability and the fact that only an energy policy can provide stability. This strategy is doomed to failure.

A stable energy policy system has three major components:

1. A policy *sector*, which provides the conceptual framework around which policy is made;
2. A policy *community*, which consists of those parties-at-interest that collectively make policy; and
3. A set of policy *decision norms*, which govern and guide the process by which the community makes policy.

The chapters that follow expand on these components and offer recommendations about what must be done to develop a stable policy system in the United States. The remaining chapters in part one trace American energy policy as it evolved in the post–World War II period to 1973. The basic nature of stable policy systems is outlined (chapter 2), and the development of energy policy in the years before the embargo is analyzed (chapter 3). Part two provides an assessment of the range of participants in energy policy after the embargo (chapter 4), discusses the range of options in the energy sector (chapters 5–7), and gives an overview of policymaking for energy between 1973 and 1980 (chapters 8–10). Part three responds to the issues raised by energy policy since 1980. Chapter 11 analyzes the effect on energy policy of the free-market orientation of the Reagan administration. In part four, chapter 12 draws the book's conclusions, and chapter 13 makes recommendations for the development of a stable energy policy system.

CHAPTER 2

The Policy System

> *The United States is in the midst of a well-publicized but very real energy crisis. The crux of this crisis lies in the contradiction between economic, political, and technological realities and our policymakers' inappropriate responses.*
> —Richard Mancke, *The Failure of U.S. Energy Policy*, 1974

The failure of the American policymaking system to anticipate and respond to the changing facts of energy that led to the energy crisis was a nearly inevitable consequence of the way policy is made in this country. The system was organized to make and manage policy for five individual fuels. As scarcity was perceived to be replacing abundance, however, the conviction grew that energy should be managed in a more comprehensive fashion, rather than by fuels. In 1973 the focus of policymaking moved from problem solving in five stable fuel policy systems to a search for issue resolution by the president and Congress (see figure 2.1). While it was generally agreed that the traditional arrangements for controlling energy no longer worked, there was no agreement on what new arrangements should replace them. Only after an issue-resolution process had occurred would it be possible to reestablish a stable problem-solving policy process for energy.

Most policymaking in the United States involves *problem solving* by stable, semiautonomous policy systems. Problem solving lies, by definition, in the domain of expertise, where the application of knowledge and information is seen as the route to policy solutions (Coates, 1979, p. 29). The stability of this country has derived from the capacity to achieve most public-policy goals through problem-solving processes.

Policymaking involving *issue resolution,* on the other hand, occurs at the presidential-congressional level and is the domain of politics, which is characterized by "conflict among or between objectives, goals, customs, plans, activities, or stake-

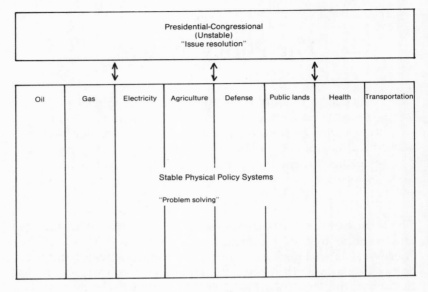

Fig. 2.1. Two policy contexts.

holders" (Coates, 1979, p. 29). Issue resolution is required to deal with crises. A crisis exists when the "mechanisms of government and their organization and structure are obsolete to a degree that has engendered a fundamental incompetence to deal with many of the new issues which the nation faces" (Coates, 1979, p. 29). Such was the energy crisis.

Most domestic policymaking has as its purpose the governance of "things in the world," that is, physical phenomena. Most policy options, then, are defined or bounded by what it is physically possible to do. A major difficulty faced by executive and legislative decision makers in this country after the oil disruption was the lack of consensus about what physical activities should be included under the label "energy" and which of these activities were in fact available options.

Uncertainty about which physical options were available was a key barrier to rapid issue resolution. To appreciate the importance of this uncertainty, it is necessary to review the general character of United States policymaking. The starting point is to note that *policy* always is preceded by a descriptor. It is common to divide policy into *foreign* and *domestic* cate-

22

gories. For both constitutional and practical reasons, foreign policy normally involves much more continuous participation by the president than is the case with domestic policy. Domestic policy includes a massive collection of fragmented policies organized around distinctive sets of substantive activities. To participate effectively in domestic policymaking requires understanding of those substantive arenas. In general, presidents involve themselves in domestic policy concerns only when something serious goes wrong—when something occurs that has broad, negative national consequences. The development of a domestic energy shortage was that kind of situation.

It is useful to divide domestic policy into *social policy* and *physical policy.* Social policy subsumes activities such as civil rights and welfare. Social policy generally involves delivery of services by government directly to clients or, alternatively, efforts by government to influence how various communities interact with each other. Physical policy includes such activities as agriculture and energy. Through physical policy government acts to control or influence distinctive sets of physical phenomena—things in the world—in such a way that those phenomena have the desired impacts on people. In the case of social policy the government interacts directly with individuals and groups. In the case of physical policy the government affects society by influencing things.

It is important to emphasize that the history of domestic policymaking in the United States is primarily a history of physical policy. Until the New Deal most domestic policy was focused on physical phenomena. The nation had distinctive policies for agriculture, public lands, transportation, and so on. Within these broader physical policy categories were programs for more narrowly defined activities. For example, within agriculture specific policies were formulated for corn and for cotton. Similarly, separate policies existed for petroleum, natural gas, coal, nuclear power, and electricity. The key is that public policy in the United States evolved primarily to manage things, not people. When the steam locomotive was developed, it was not long before policy was put in place to manage railroads.

In general, policies for major new physical activities are formulated when the new activities create issues that must be resolved by the president and the Congress. Once issues have been resolved, new policy systems created, and new policies formulated, responsibility for the evolution and implementation of those policies is transferred to semiautonomous, stable policy systems, where problem solving is the policymaking style. When something goes wrong with the way things are managed in these stable systems and new issues appear, the president and the Congress again become actively involved.

Two points need to be restated. First, *physical policy starts from things in the world, and the nature of those things establishes the boundaries of policy and defines the available policy options.* Second, *once the need for policy is established, the policymaking process is played out in two very different ways: issue resolution (at the presidential-congressional level) and problem solving (within semiautonomous policy systems).*

The establishment of major new policy requires issue resolution at the highest levels of decision making, the president and the Congress. Issue resolution is fraught with conflict. It is the conflict-ridden, issue-resolution activities of the executive and legislative branches of the federal government that are the focus of media attention and analysis by political scientists.

Policy stability usually is established for things only after the president and Congress have resolved the four kinds of issues outlined for energy in chapter one: (1) definition of the goals to be pursued, (2) definition of the things to be influenced or controlled, (3) definition of the instruments to be used, and (4) definition of the appropriate political, managerial, or organizational mechanisms to be created or modified. When the president and the Congress have resolved these issues, the conditions exist for meeting the policy needs of physical activities through a process of problem solving. That is, when issues are successfully resolved, a consensus is established that makes it possible for a different kind of policy process to take over—problem solving.

The making of policy by problem solving typically is managed within much more limited systems, and these stable systems operate under very different rules with a much more limited range of participants. Specifically, the norm in the United States is for policy to occur by evolution, with decision making by consensus. Moreover, policymaking of this kind takes place within physically bounded, self-contained systems away from the mass-media limelight. In short, the United States has developed a system that allows vested interests to make a great deal of policy for themselves.

Such arrangements generally allow those with a major interest in a given set of physical activities to seek high-level (presidential-congressional) action to legitimize policy changes around which a consensus already is formed. In the parlance of Washington, D.C., there is no "federal government"; rather there are multitudes of governments. It is probably more accurate to say that there is no "federal policy"; rather there are many federal policies produced by many policy systems. This condition is not a malady; quite the contrary. Given the fragmented character of American society, such policymaking accounts for much of the nation's political stability.

This two-level policymaking system began in the earliest days of the nation. The unique character of American policymaking derived from five historically distinctive circumstances: (1) the historical and intellectual factors that led to the American Revolution and to the evolution of the American Constitution, (2) the dominant focus of Americans on things material, (3) the impact of wave after wave of immigration into the United States, (4) the abundance of natural resources, and (5) the key role of scientific and technological information and activities.

The formative period of policy development in the United States was dominated by persons who shared a view that the traditional conception of sovereignty and the apparatus of hereditary rule that characterized European governments should be rejected. For these concepts the founding fathers substituted "the idea that the people had the right by rational and experimental processes to build their governmental institutions to suit themselves" (Price, 1954, p. 5). The scholarly literature is filled with discussions of these early efforts to protect against traditional notions of sovereignty by separating the powers of government, putting in checks and balances, and building a federal system in which power was divided between a central government and state governments.

From today's vantage point, however, these attempts at dividing and fragmenting power and sovereignty appear less distinctive than the pattern of building policy systems that were physically oriented. To a striking degree American policymaking has been unaffected by ideology or social class; substantive experience and expertise have been the engines driving policy. Clearly the role of science and technology has been much more important in framing American policy choices and in defining policy options than has any body of value theory.

The materialism of Americans, the physical abundance of the North American continent, and the influence of massive numbers of immigrants made the United States a particularly fertile environment for physically oriented, expertise-dominated policymaking. However these developments are viewed, no theme pervades the observations of the American experience like the focus on materialism. In its initial period the nation was faced with overcoming a hostile natural environment and, following that stage, with extracting from that environment the materials that allowed the new nation to become rich and powerful. The importance of science and technology in the United States derived from the need first to understand and catalog the natural environment and then to extract materials and utilize them to satisfy desires or to escape adverse risks and consequences (see Noble, 1977). The development of policy systems around integrated, self-contained physical activities made policy dependent upon those bodies of scientific and technological information that described and provided understanding of the physical world.

The materialistic orientation of Americans clearly would not have been possible but for the rich natural abundance of this continent. Initially the nation had a seemingly endless western frontier. Once the geographical frontier was gone and the country became continental, it appeared to have a nearly endless supply of resources, including energy. The coupling of materialism with natural abundance provided the rationale and magnet for immigration. For those who were already here, immigrants supplied the human resources required to expand the national economy by developing, extracting, and utilizing its resources. For the newcomers the message of material opportunity obviously was powerful, and although some historians have argued that the beacon of America was political, material abundance surely was a major component of this country's attractiveness.

If one considers for a moment the striking character of the evolution of policymaking in the United States, the importance of physically focused policy systems to political stability

becomes evident. It is almost a cliché among students of political systems and government that the ultimate obligation of
such systems is to survive. The ideal achievement of political
systems is relative stability, or, as Easton (1965, p. 84) has put
it, "persistence with and through change."

If students of political systems were asked how best to set
up a stable system, most of them would include three major
cautionaries: first, be sure not to fragment power or sovereignty unduly; second, have a homogeneous population; third,
have a populace whose expectations are reasonable and who
do not put extreme demands on government.

The United States started with a self-conscious attempt to
fragment power and authority. It then introduced immigrants
who made up a heterogeneous population. Finally, it encouraged in that population unbounded expectations. In sum, what
took place in the United States was the diametric opposite of
the normal formula for stability. When these circumstances
are combined with other facets of American development, such
as the construction of the country from thirteen colonies with
no special affinity for each other, all the ingredients for political and social stress and cleavage appear to be present. The
genius of the ad hoc creation and maintenance of policy in
the United States can be appreciated by noting that, despite
all the factors tending toward disequilibrium, United States
history has been one of impressive social and political stability.

What happened in the energy field, then, was perfectly consistent with the historical pattern of policymaking. Policy was
made around distinctive sets of physical activities (the five
fuels), the policymakers being those who had a vested interest
in a specific fuel. When the physical facts of energy changed
so that the policymaking structures were no longer capable
of managing those facts, instability was introduced, and the
president and Congress were faced with formulating a new set
of actions and developing an entirely new system. Once a new
national consensus had been generated and a new system was
in place, the management of energy could be turned over to
those participants who had a vested interest in the physical
activities of energy. It was this process of developing consensus that kept the executive and legislative policymakers busy
after the embargo.

Thus both the source of the energy crisis and the answer to it can be understood only if one understands how stable policy is made. It is then necessary to investigate the structure and processes of the relatively autonomous, stable policy systems that manage physical activities in this country.

STABLE PHYSICAL POLICY SYSTEMS

Three major elements are central to understanding the essential characteristics of the stable policy systems that have made and managed physical policy in the United States: the policy sector, the policy community, and the decision-making norms. The policy sector provides the conceptual framework around which policy is made. The policy community consists of those actors and interests who collectively make policy. The decision-making norms are those rules that govern and guide the process by which the community makes policy. These elements are illustrated in figure 2.2.

The Policy Sector

As noted above, policy is always preceded by a descriptor—the distinctive set or sets of physical activities that are to be influenced or controlled by the actions of government. The boundaries of policy sectors are defined by what is physically possible or what the participants in the system believe to be physically possible. This point can be illustrated by noting that it was possible to decide to put a man on the moon and bring him back alive but it is not possible to decide to build an antigravity machine. The difference between these two policy options is that, in the first case, natural laws permit putting a man on the moon. Alternatively, those same natural laws mandate that an antigravity machine cannot be built. At a minimum, then, the boundaries of policy sectors are defined by natural laws.

Since science is the source for understanding these laws and is a dynamic activity, the boundaries of policy sectors can, in fact must, change as comprehension of nature changes. It is clear that an understanding of nature through science establishes a set of constraints for the policy options that are

28

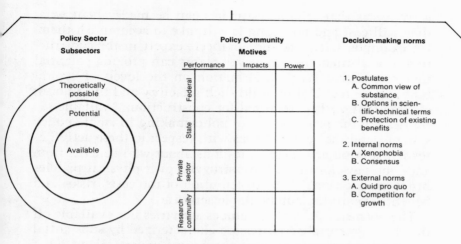

Fig. 2.2 Elements of stable physical policy systems.

available in any sector. Similarly, technological capabilities establish constraints on what policy options are available.

It is equally important to note that changes in understanding coming from either scientific discoveries or technological developments can be a major force driving policy change (Boorstin, 1978). The continuing process of technological evolution and scientific discovery is an ever-present initiator of policy modification in the technological American society.

The most successful stable policy systems have integrated this expectation of physical change into their routine procedures and handle them in a problem-solving manner. Policy systems do this by refining the concept of the policy sector. That refinement involves dividing the sector into three subsectors. In figure 2.2 all the activities that fall within a given policy sector are enclosed in a circle. To illustrate the role of scientific and technological change in policymaking, the sector circle is divided as follows: the innermost circle is labeled the available subsector. The second, or intermediate, circle is called the potential subsector. The outer circle is termed the theoretically possible subsector.

The *available subsector* includes those technologies, processes, and physical activities that can be identified as "off the shelf." The major distinguishing characteristic of available

29

activities is that reliable estimates can be made about what they will cost and how long it will take to accomplish them. For example, scientists and engineers expert in the construction of coal-fired electric power plants can provide potential investors with reliable information on the levels of capital that will be required to build such a facility and the amount of time that will be required for construction. Available activities are the primary foci of policymaking in stable policy systems, and the central issues with respect to these activities revolve around answering questions about who benefits from and who pays for them. Most struggles over physical policies hinge on questions of the distribution of the costs, risks, and benefits of carrying out available activities.

The *potential subsector* includes activities not available off the shelf. Potential activities are characterized by substantial disagreement among scientific-technical experts about what they will cost and how long it will take to carry them through to conclusion. This disagreement is distinctive because the many technical uncertainties "leave considerable leeway for conflicting interpretations" (Nelkin, 1981, p. 12), and therefore the risks, costs, and benefits of these activities are ambiguous.

Normally, activities that fall in the potential subsector have been tested on a small or laboratory scale but have not been constructed and operated on a commercial scale. For example, proposals for coal-based synthetic-fuels plants are characterized by high degrees of uncertainty regarding product cost, plant reliability, and ecological risk. A very substantial portion of the energy policy debate in this country has centered on disagreements about what the policy should be in the presence of such uncertainties.

One of the major failings of the fuel policy systems in existence before 1973 was that they were incapable of managing potential activities. A common problem in this area is that some technical experts argue that activities are available while others maintain that they are merely potential. As Mazur has pointed out (1981, pp. 10–33), such disagreements among experts originate in differing interpretations of data. The disputes present complex obstacles to policymaking, especially when expertise is mobilized in an advocacy fashion and polarization takes place in the debate about potential technological alternatives, as occurred after the embargo.

The *theoretically possible subsector* includes those activities in which scientific and technical specialists agree that options are not yet available but are theoretically feasible. Theoretically possible activities, by general consensus, cannot be carried out within any predictable time frame or at any predictable risk or cost. An example of this kind of activity is the generation of electricity from fusion.

A key ingredient of stability in a policy system is the level of agreement among scientific and technological parties-at-interest regarding where an activity falls—whether in the available, potential, or theoretically possible subsector. Any significant disagreement over the location of these activities poses a serious challenge to the stability of the system.

On the other hand, stable policy systems are well organized to deal with policy changes when the movement of options from one subsector to another is orderly and accompanied by reasonable levels of consensus within the expert community. The use of physically interrelated activities as the framework for policymaking therefore creates a very important role for the expert (see Nowotny, 1981).

The Policy Community

The second major component of physical policy systems is the policy community, which consists of those actors who have an ongoing interest (or a perceived interest) in what actions take place within the policy sector. Stable policy communities contain actors who agree about which activities are available, potential, and theoretically possible. Stability also requires a consensus regarding which parties-at-interest have a right to participate in the community. In a stable policy community participation is viewed as a right for those actors who demonstrate both a willingness and a capacity to play an ongoing role. The emphasis is on the *continuity* of involvement. Actors who participate intermittently are generally perceived as disruptive and as threats to long-term stability because they will likely be substantively ignorant. Similarly, participants without a clear vested interest—a stake—in the activities of the system are viewed as potentially destabilizing influences. Something of the complexity of the participants in stable systems can be appreciated by discussing their motivations for involvement and their organizational location.

The *motivations* that drive actors and interests to participate in policy communities can be divided into three categories: performance, impact, and power (Rycroft and Szyliowicz, 1980), as illustrated in figure 2.2. The performance motive has to do with designing physical policies so that they solve substantive problems. Performance typically is defined in terms of a rational-choice orientation in which productivity and efficiency are the dominant values, faith in the technological fix as a solution to almost any problem is virtually unchallenged, and politics is viewed as a corollary of ignorance, to be overcome by the application of specialized knowledge (Lindberg, 1977).

The impact motive focuses on the desire either to maximize the benefits generated by physical activities or to minimize the costs and risks incurred by these activities. The benefits, costs, and risks may be economic, but they also include a range of other kinds of "products," including health and safety, ecological, and aesthetic concerns.

Power, or control, as a motivating factor for participants emphasizes the traditional question of political influence in a society. At its root the power orientation stresses the attempt to gain access to the levers of control in a policy system— to exert authority over decisions. Of course, there are major areas of overlap among these categories, and actors typically participate for a combination of motives.

The members of policy communities who participate predominantly for performance reasons are the scientific-technological actors. In the American policymaking context it is expected that scientists and engineers, working in their professional spheres, will develop most of the basic ideas to which stable systems ultimately will turn to deal with problem solving. Price (1954, p. 27) provided a perspective on this rather distinctive American dependence on expertise by noting that "to the foreign visitor the most striking phenomenon in American government is likely to be the extent to which private research institutions play a continuous role in the development of government administration."

Particular attention needs to be focused on the performance motivation. Scientific and engineering interests have a drive that transcends any particular goals and objectives of organizations within which scientists and engineers may be located. It is the recognition of this commitment to performance that leads other members of the policy community to accept scientific and technical definitions of the boundaries of the policy sector. Thus in a stable policy system the policy community can rely on scientific-technical participants to define those activities which are available with the expectation that these definitions will be relatively noncontroversial, having been based on a shared consensus founded in a common body of theory, data, and experience. If for some reason this definitional role is not carried out, very destructive seeds of instability are likely to be planted in the policy system. The origins of most policies in the United States can be traced to discussions that have taken place among various scientific and technical participants within a given physical policy system. Policies that evolve from a segment of the community that has achieved a consensus built upon science and technology are likely to achieve and maintain stability.

Although performance is the motivation for most expertise in stable physical policy systems, it is concern about the impacts of activities within the policy sector that mobilizes most of the actors and parties-at-interest. The continuing participants in a policy community motivated by impact concerns have historically been of two kinds. First are those parties-at-interest whose economic well-being is intimately tied to actions in the policy sector. Traditionally, the most active policy participants are those private-sector persons or organizations that are involved in carrying out the production process in a policy sector. Clearly oil firms participated on a continuing basis in the oil policy system because they had the most direct vested interest in the activities that were undertaken.

The second group of impact-motivated participants is made up of those public-sector organizations or agencies, particularly regulatory bodies, who have legally defined roles in the management of activities in the policy sector. Before the development of more broadly based consumer and environmental interests, policy communities had a tendency to be relatively bipolar. That is, they involved on the one hand a group of government agencies and on the other hand a group of participating business interests. This circumstance raises an important point. For most of American history there has been little concern with second-order impacts. The environmental movement, however, focused a great deal of interest and energy on impacts that went far beyond the classical notions of formal-legal responsibility of government structures and of economic impact within the private sector. In short, the focus upon ecological externalities in the 1960s resulted in new sets of actors seeking to participate in a broad range of physical policy systems. This expansion of the range of impacts to be considered in the policy systems greatly complicated the policy-making process. The expansion of impact concerns certainly made energy decision making much more complex and controversial.

One point deserves special emphasis: not all affected actors and interests participate in policymaking within any physical policy system. Moreover, some participants may not be greatly affected by activities in the physical policy system. Note, for example, that the primary actors in the electric power policy community were the state regulatory agencies and the electric utilities. The consumers of electricity obviously were affected in an ongoing manner by policy decisions that modified electric power activities, but they were not and are not usually participants in the community. On the other hand, the nuclear power arena offered an especially attractive place for a number of actors to participate in a "vicarious" way—antinuclear power protests have every bit as much to do with resistance to weapons proliferation and to the perceived centralization of scientific and technological decision making as they do with any energy issue or problem (Landsberg, 1980).

The third category of motives for participation in policy communities is power. Power-motivated participants are predominantly concerned with either protecting or expanding their roles in policymaking. The power motivation is particularly evident in government participants in policy communities. For example, under the federal system state governments frequently contend with the federal government for the right to influence particular physical policy systems. Obviously one of the reasons for the struggle between state interests and federal interests is that they may represent different sets of affected parties. Beyond that, however, the traditional federal-state struggle plays an important role, state actors being concerned about the protection of states' rights (Lamm, 1976).

In the same way the "turf" concerns of both congressional

committees and federal executive agencies are a pervasive factor in policymaking communities. In the initial efforts to formulate a national energy policy, congressional committees frequently manifested less concern about the performance and impacts of energy than about their own turf. That is, their motive was to gain a continuing role in the new energy policy system. The same phenomenon is evident with regard to the federal executive agencies, which are strongly motivated to protect and expand their authority. It is important to understand that desire for power motivates not only individual participants in policy communities but also bureaucratic (and organizational) actors. Congressional committee chairmen have reputations for protecting and seeking to expand their areas of responsibility. There is within the Congress a folklore about the ability of skilled congressional committee chairmen to enlarge their domain or at least maintain it in periods of instability.

A more pervasive illustration of the power drive, however, is seen in bureaucratic parties-at-interest. Policy communities evolve professional groups of managers in the executive branch who take on the role of protecting their organizations. The motives for bureaucratic concern about organizational survival are complex; it is common to argue that a major factor is the desire of public servants to protect their jobs. The self-interest explanation does not take into account other factors, however. For example, participants in large bureaucracies tend to take on the identity of the organization. Bureaucrats themselves and analysts of bureaucracy repeatedly note the phenomenon, and there seems to be a rather constant pressure for bureaucracy to expand its turf (Wilson, 1975).

In almost all areas of policymaking the key to active participation is organization by the concerned parties. Only organized actors and interests are able to meet the requirement for continuous interaction because only organized interests can stay abreast of scientific-technological changes. Policy communities are informal systems that link Washington with other levels of our society. Affected actors and parties-at-interest who are involved on a continuing basis are those who have the ability or the concern that leads them to organize and to become ongoing, stable participants.

Although motives are important in linking the actors in

35

policy communities, regardless of where those actors may be institutionally situated, *location* is itself a significant variable. Parties-at-interest in policy communities may come from any of four institutional settings: the federal government, state governments, the private sector, and nonprofit research organizations, usually based in universities. Actors and interests from any of these settings may be driven to participate in the policy community for any combination of the motives we have just discussed.

Although the mix of motivation and institutional location varies from one physical policy system to another, some general observations can be made. First, the performance-motivated participants—scientists and engineers—are increasingly located in all four institutional settings, though their concentration tends to be higher in universities and in the private sector. Impact-oriented participants tend to be more heavily concentrated in the private sector, though the consequences of policy actions are of concern in each of the institutional settings. Similarly, the power motive is more evident in both federal and state governments, but again is manifested in all four sets of institutions.

It is possible at this point to draw some tentative conclusions about the structure of stable policy systems and especially of policy communities in the United States. The starting point must be to reemphasize the importance of the *policy sectors*—distinctive sets of physical activities—in forming the framework for physical policy. The policy communities that organize around and make policy for the activities in these individual sectors consist of participants who have an ongoing vested interest in the physical activities.

A compelling characteristic of actors in policy communities is their willingness and ability to organize to ensure themselves continuous access and involvement. Participants in policy communities communicate with each other constantly. A primary medium of that communication is the set of scientific-engineering experts who are concerned with performance. Policy communities can be viewed as information networks that deal primarily with substantive information and operate with little regard for organizational or hierarchical position. Stable policy communities consist of actors and interests who are continuously engaged in obtaining, digesting, exchanging, or testing information. Thus a very important part of the process by which these policy communities function is communication—through word of mouth, discussions in the workplace, meetings sponsored by professional societies, and numerous informal ways.

Decision-making Norms

Decision-making norms, the rules that govern how policy communities evolve their choices, are the final element in the stable physical policy system. While government is the promulgator of policy in stable systems, this action normally occurs only after policies have been formulated by a process of consensus. The agricultural policy community illustrates the mix of institutional participants in a policy community and demonstrates how a diverse group of actors is held together by communications networks and similar modes of decision making. Members of the agricultural policy community include the agriculture committees of both the House and the Senate, the Department of Agriculture in the executive branch, and a large number of private-sector organizations, including seed-corn companies, grain firms, farm-machinery manufacturers, and the American Farm Bureau Federation and the Farmer's Union, two organizations representing different sets of agricultural interests. Finally, a very significant role is played by employees of the agricultural-research stations and those in the land-grant universities, such as Texas A&M University or Iowa State University.

If one conceives of the congressional committee as the top and the individual farmer involved in a specific policy sector activity as the base, the policy community can be seen as running from top to bottom resting on a communications network that is rather similar to the information flow pattern in a small rural society. Physically oriented stable policy communities have very effective informal information networks, and they exercise very real power through the use of this network. Power flows in part from the community's superior access to information. Informal policy communities are thus held together by a common reference point—the policy sector—and by a group of operating rules, or choice procedures. All the agricultural interests mentioned above share the underlying assumptions, values, and norms by which policy decisions are made.

There are two fundamental kinds of decision-making norms: internal and external norms. The interaction among participants *within* the policy-making community is governed by *internal norms*. Stability requires three areas of agreement. First, all the community members must share the same definition of the policy sector and subsectors. They must agree on what activities are available, potential, and theoretically possible. Second, the members of the community must frame policy alternatives in scientific, technical, or administrative terms—something close to what Braybrooke and Lindblom (1970, pp. 78–79) termed "synoptic" decision making. In other words, policy options must be developed from an understanding of the physical activities, and policy must be seen as problem solving. Third, members must agree that the highest-priority goal is to protect the rights, interests, and benefits that already exist in the policy system (or, on the other side of the coin, to protect the policy community from additional costs or risks).

Starting from these shared commitments, all participants in a stable community have one fundamental obligation: they must make every effort to exclude nonmembers. Therefore, members cannot go outside the community seeking allies in an effort to achieve short-term objectives. This norm is labeled *xenophobia.*

The fear of outsiders proceeds from two concerns. On the one hand, outsiders may not share the commitment to protect existing benefits or reduce additional risks. Moreover, nonmembers may not share the same perception of the policy sector. Outsiders frequently want to introduce sudden, difficult-to-manage changes in the way the physical activities are carried out. In the fuel policy systems, for example, outsiders wanted to resolve supply-demand imbalances through conservation, while insiders were committed primarily to dealing with energy problems and issues by increasing supplies.

The second internal decision-making norm of stable policy communities is one already discussed at some length—*consensus.* This does not mean that every participant agrees across the board all the time. Rather, it means that the decisions that are made are sufficiently acceptable to all the parties-at-interest that they are at least willing to go along. In consensus decision making, choices are made in such a way that actors do not perceive the consequences as so damaging that they must go outside the community to seek support to protect their interests. In Landsberg's words (1980, p. 84), consensus building requires "a willingness to moderate demand for adoption of one's entire agenda." Consensus decision making in stable systems almost always means that decisions are made in an incremental, piecemeal fashion. Moreover, this incrementalism rests very heavily on the existence of a scientific-technological definition of policy choices. Developing policy choices from a technical basis usefully plays down value differences and emphasizes the assumption of a basic consensus. By defining alternatives in technocratic terms, the policy debate is framed under the essential assumption that if more information were available or broader understanding were possible everyone in the community would agree. Further, the informal communications system maintained among scientific and technical participants allows the community to play down normative differences or at least to blur them sufficiently so that polarization does not occur.

In sum, the great body of policy debate and interaction within stable policy systems takes place in a context in which the competing actors and interests believe that they share the same policy objectives as long as they can negotiate what they define as administrative or technical issues. Such debates proceed in an incremental way, but with the goal of greater rationality and comprehensiveness always just over the horizon.

External norms govern the ways in which the various policy-making communities interact with each other. Just as stability is dependent upon decision-making rules accepted within the community, so it is in part a function of agreed-upon procedures for interaction among policy systems. Two norms dominate relationships among stable policy systems: the *quid pro quo* and *competition for growth* rules.

Throughout the American political system policy communities share a commitment to the *quid pro quo* rule. Simply stated, the rule is: "We commit ourselves not to interfere with your policy system if you do not intervene in ours." This "rule of the game" is the inevitable derivative of a political system that allows actors and interests who have a perceived continuing vested interest in a physical area of activity to make policy for themselves in relative isolation.

The essential commitment here is that members of any policy community will never take actions aimed at appropriating rights or benefits that are held by another policy system. For example, when government revenues are committed to some policy system, members of other policy communities will not try to appropriate them. Similarly, when policy systems have preestablished legal rights or advantages, participants in other policy systems will not try to reverse them.

The *competition for growth* rule flows from the historic American assumption of infinite abundance. This assumption provides a relief valve for the pressures for growth or expansion within policy systems. The rule says that it is acceptable to compete for that portion of the economic pie that is added each year by real economic growth. In the same vein, it is acceptable for each policy system to compete for additional legal, regulatory, and budgetary assistance or tax advantages, as long as that competition does not reduce the advantages held by another policy system. The American political system is pervaded by the message that no one has to give up anything to permit others to improve their position. This was the case with the fuel policy systems. Political debate in the United States consistently has rejected the notion that it is necessary to redistribute wealth even while policy was doing it, because the overall quantity of wealth has been expected to grow. The competition for growth rule is an essential ingredient for stability.

It must be emphasized that the enforcing mechanism for both external-norm rules is the assurance that their violation will be followed by retaliation by the affected policy system. Stated crudely, "If you interfere in the affairs of our policy system, we will do the same thing to you." Ultimately the enforcing mechanism is mutual reluctance to introduce *instability* into the existing system.

SOURCES OF INSTABILITY

Instability is the ever-present fear that binds together the actors in stable policy communities. The threat of instability makes these systems inherently conservative and incremental in their operation. Most stable policy systems are subjected to intermittent pressures by outsiders for change. As long as those pressures do not become constant, physical policy systems generally are quite successful in fending them off. In typical situations, stable policy systems simply do not react or respond to intermittent pressure.

On the other hand, with sustained, organized efforts, outside actors and interests usually develop the skills to force change in even the most stable system. Under constant pressure two things can happen. One is the onset of a period of short-term instability during which the stable system integrates the new interests or coopts them. If this process is successful, stability is reestablished, but the policy community now includes new actors, and the substance of policy is likely to be modified. When stable communities are not successful in this integration effort, it is usually necessary for the president and Congress either to reformulate the system or to create a new one. The latter is what took place in the area of energy.

A number of events can trigger sustained pressure for change in policy systems and thus create instability. Such events include (1) catastrophes, (2) changes in social values, (3) a sudden scarcity of a natural resource, (4) unique actions by individuals, (5) major scientific or technological advances, (6) action in one policy system that produces inadvertent consequences for another, and (7) perceptions of basic deficiencies in the social system. Any of these events can generate demands by previously external interests to participate in a stable policy system.

Catastrophic events can have the effect of mobilizing previously nonparticipating actors and interests by demonstrating that existing policy systems are not protecting values important to them. An example is the Santa Barbara oil spill, a catastrophic event that served as a catalyst for a nationwide mobilization of parties-at-interest who insisted that major modifications had to be made in the management of offshore petroleum and natural-gas activities (Kash et al., 1973). Similarly, the Three Mile Island nuclear accident brought additional support to those antinuclear actors who were already opposing the development of nuclear power (Gilinsky, 1980). The opportunity provided by catastrophic events to mobilize demands by outsiders for access to stable policy systems is

41

enhanced significantly by modern-day mass-media techniques.

Periodically, in the evolution of policy systems, fundamental changes in social values take place. The environmental movement, and the high priority it attached to ecological well-being, is perhaps the most significant example of such a change in recent times. When such shifts occur, stable policy systems may find that their goals and objectives are in conflict with the modified values. Moreover, the inherent conservatism of stable systems required by their commitment to consensus decision making and therefore to incremental change usually makes it difficult for the policy system to incorporate new values. The environmental movement demanded that a number of policy systems, including the fuel systems, integrate new actors and modify policies to respond to shifts in values.

Natural-resource scarcity results from changes in the physical environment. Such scarcity can trigger demands from a broad spectrum of actors and interests to participate in the policy systems that manage the resources. In the fuel policy systems these demands were accompanied by violations of the quid pro quo and growth-competition rules. As long as truckers and farmers were supplied with ample petroleum products at low cost, they subscribed to the rules of the game. But when oil became scarce and costly, other affected policy systems sought to intervene in the management of the resource: the transportation and agricultural systems began to intervene in oil policy.

American history is characterized by instances in which individuals have been able to mobilize previously noninvolved actors to demand participatory rights. Perhaps the most widely recognized individual of this kind in recent years is Ralph Nader. The publication of Nader's book *Unsafe at Any Speed,* a critique of the auto industry, made him a national figure. Nader's unique ability to communicate across a broad spectrum of society through skillful utilization of the mass media led to major interventions in the transportation system. The actions of a key individual thus had the effect of opening a system to a much broader based organization of consumers. It should be noted that Nader perceived, with a clarity that was rare until his involvement, how important it was to be equipped with data about physical activities that occur within the policy sector. A major reason for Nader's success was that he challenged the stable policy system with technical-scientific information. In recent years the same strategy has been applied to energy, with limited success, by various Nader organizations and a host of other public-interest lobbies (McFarland, 1976).

Scientific and technological breakthroughs have repeatedly inspired new actors to mobilize in demanding greater participation in stable systems. More broadly, however, a new scientific or technological advance may create stresses that have much greater implications than the introduction of new parties-at-interest. Perhaps the most dramatic example of what can occur with scientific and technological breakthroughs is seen in the communications industry. New technology changed the character of the interstate communications policy system, created competitors for the Bell system's domination of the market, and led to the breakup of Bell. But the revolution has gone much further, producing competitors for the major television networks and significant changes in the way information is accessed, controlled, and utilized by the consumer. Ultimately, the entire social system will be affected.

On occasion new participants are introduced into stable policy systems as a result of decisions made in other policy systems. An example is the decision to establish the food-stamp program. This action resulted from choices made primarily within the welfare policy system; however, responsibility for the handling of food stamps was given to the U.S. Department of Agriculture. That decision brought demands by food-stamp recipients to participate in the agricultural policy community. Not surprisingly, the participation of food-stamp consumers in the agricultural system has been the cause of some instability. Before the initiation of the food-stamp program the agricultural system was oriented toward the objectives of the producers and processors of food. After the introduction of these new actors the system had to deal with new demands and expectations, to say nothing of different value sets. All of this has made consensus building much more difficult.

The final category of circumstances that can trigger sustained pressure for change includes those occasions in which fundamental deficiencies are perceived in the conceptual-organizational-managerial behavior of the social system. The two most easily identifiable occasions are economic depression and war.

43

In these events few stable policy systems are unaffected. The Great Depression and Franklin Delano Roosevelt's "New Deal" represent an example. Some believe the election of President Reagan may represent a similar case. Clearly, the Reagan administration saw its mandate as requiring a restructuring of the role of government in society. Certainly presidents have a capacity to destabilize policy systems that is unmatched by any other actors.

Although any one of the above situations may lead to demands by outside actors and interests for entrance into stable systems, more commonly some combination of circumstances triggers instability. Such was the case with regard to each of the five fuel policy systems in existence at the time of the oil embargo. After 1973 new actors and interests demanded the right to participate. Elements of each of the seven sets of circumstances outlined above can be seen in this development.

The following illustrations indicate something of what was happening at the time of the embargo. The oil-pollution accident in Santa Barbara was a primary reason for the rise of the environmental movement in this country, but the process took place in the midst of changing values with regard to the costs and risks of ecological degradation. Perceptions of energy scarcity were magnified by the limits-to-growth ideology that had become widely diffused. Actions by important individuals like Rachael Carson; the emergence of a range of alternative clean-energy technologies (particularly solar power); the inadvertent linking of energy, economic, and environmental concerns; and a number of perceived deficiencies in the social system (the most striking of which was the rapid increase in inflation) generated massive pressures for change in the way the nation managed energy.

Thus a wide variety of sources of policy instability converged at the time of the embargo, with the consequence that energy was placed on the presidential-congressional agenda where it was soon to occupy a high place. The circumstances that had allowed the country to manage energy in a stable condition for many years before 1973 made it impossible for the nation to respond rapidly to the energy crisis. Stability with regard to the energy system would be reestablished only when a new policy system could be constructed for the energy field as a whole, supplanting the no-longer-stable policies for specific fuels.

In the next chapter we investigate the policies that were in effect for the five fuels at the time of the embargo. It is important to remember that all these policies rested on the assumption that the future would call only for the continuation of a capability to manage surplus.

CHAPTER 3

Energy Policy Before 1973

The Arab oil embargo, while it lasted, made us keenly aware that in the twentieth century America, a fourth essential has been added to the age-old necessities of life. Besides food, clothing, and shelter, we must have energy. It is an integral part of the nation's life support system. And we can no longer expect to get it with so little trouble and expense as we did in the recent past. — Ford Foundation Energy Policy Project, *A Time to Choose*, 1974

If things had worked out as envisioned, the United States would now have entered its fourth energy era, the nuclear era. Thus far the nation has moved through three energy eras, as illustrated in figure 3.1. The first was the era of wood. The second, beginning in the nineteenth century and continuing to World War II, was the era of coal. The third, in the postwar period, was dominated by petroleum (or, more precisely, oil and natural gas). During the 1950s and 1960s it was expected that cheap nuclear power would be the replacement for oil and gas. This expectation assumed a continuation of the relatively simple transitions between energy eras. Always before, more convenient fuels had become available to replace older, more expensive fuels. So it was to be with nuclear energy.

Replacement of one fuel with another was the result not of conscious design or national policy but of simple economic choices. In the 1970s, however, the traditional process of energy substitution was disrupted. Two arguments have been advanced about the role of national policy in causing the energy crisis. One is that government actions led directly to the crisis; the other is that government inaction caused the crisis. To determine the validity of these arguments, it is necessary to consider the fuel policies that were in effect at the time of the embargo.

Although there was no national energy policy before the em-

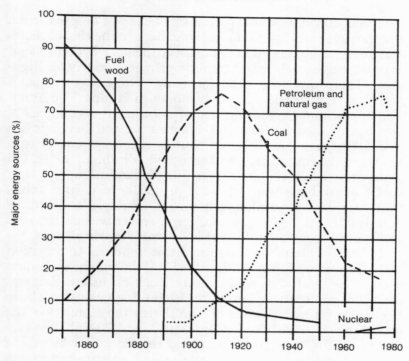

Fig. 3.1. United States patterns of energy consumption, 1860–1980.
Adapted from Lapp, 1976, p. 79.

bargo, policies for specific fuels did exist, and they shared
some common characteristics. These shared qualities flowed
from two largely unstated but universally accepted national
goals with regard to fuels: (1) the United States should have
available ever-larger quantities of energy, and (2) the unit cost
of energy should continue to decline. The policies developed
in the various fuel systems were designed to ensure abundant,
stable, low-cost energy supplies. Cheap, abundant energy was
the basis of social and political stability. It provided the foundation for economic growth and for the acceptance of common
societal rules.

The nation's commitment to the goal of ever-larger quantities of energy at ever-lower prices was pervasive. As these goals

were manifested in public policy, the focus was on production and transmission. No public policy existed in the United States that addressed energy demand. Government, whether federal or state, consistently taxed gasoline primarily to pay for highways. There was no pattern for using petroleum as a major source of tax revenues, as was common in Europe. The American approach differed from that of most member countries of the Organization for Economic Cooperation and Development (OECD) in that there was no effort to encourage energy conservation (Darmstadter, Dunkerly, and Alterman, 1977). Oil policy focused on the management of a surplus supply. Similarly, natural gas was regulated primarily to ensure stable, cheap supplies. In parallel fashion, at both the federal and the state levels, the regulation of electric power was designed to assure ever-larger quantities of lower-cost electric energy.

An event of signal importance in this trend was the creation of the Tennessee Valley Authority (TVA) during the early days of the New Deal. One of the motives behind its creation was the belief of President Franklin Delano Roosevelt that cheap electric power was a means of stimulating the economy of the region that it would serve (Davis, 1982, pp. 173–75). The success of the TVA in producing cheap electric power by taking advantage of the economies of large-scale electricity-generation plants sent a signal to all public and privately owned electric power organizations in the country. From the 1930s on, electric utilities moved rapidly in the direction of larger generating facilities that could produce power at lower and lower costs. In later years a major reason for federal support of nuclear energy development was the faith that cheap nuclear power would lead to electricity so inexpensive that it would not be necessary to have electric meters on homes (see Bupp, 1979).

Acceptance of the goal of cheap, abundant energy was almost universal. It was so easy to meet the goal that it went virtually unchallenged. Such was the situation from the end of World War II to 1973. National expectations changed in only one fundamental aspect during that period. In the late 1960s and early 1970s the standard of clean energy was added to the national calculus. No area of activity has been more deeply affected by the environmental movement than energy. The enactment of the National Environmental Policy Act (NEPA) in

1969 reflected a truly major change in social values. A clean environment became one of the fundamental expectations of society (see Rosenbaum, 1977).

Even in the decade after the oil embargo, a period in which much was made of the economic costs and energy-limiting impacts of protecting the ecological system, those calling for modification of major environmental laws were distinctly unsuccessful. For instance, although the Reagan administration came to power with substantial public support to reduce the overregulation of business, efforts to rewrite the Clean Air Act have been met with stiff legislative resistance (Lave and Omenn, 1981).

Between 1970 and 1973 a clean environment and abundant, cheap energy were seen as relatively compatible standards. When all fuels were perceived as plentiful and available at declining real costs, the answers were easy. If coal was a major contributor to pollution and a dirty environment, the response was to switch to natural gas or to find technologies that would clean up the pollution. With the real price of both natural gas and electricity declining, the cost of cleanup was viewed as so low as to be insignificant.

FUELS POLICIES

If abundant, low-cost, clean energy characterized the national expectation before 1973, what policies were in place to assure that these goals were attained? The following sections outline the policies for coal, petroleum, natural gas, electricity, and nuclear power. Public policy for and government involvement in each of these fuel systems ranged across a broad spectrum but provided for stability in all the systems except coal. David Davis (1974, p. 13) characterized fuel policies as they existed before the embargo:

Coal comes first because it is least subject to government control. It is the most private of the five fuels. Ownership is in private hands. Links to the government are minimal. No federal agency routinely regulates the price of production. The oil industry is less autonomous. While ownership is private, the linkages to federal and state governments are extensive. No agency regulates price, but a coordinated set of state commissions does regulate production. Natural gas is third. Ownership is still private, but a federal agency heavily regulates price, production, sales, and construction of the interstate industry. Ownership of electrical power companies is next. Some are private; some are public, owned by federal, state, or local governments. Still others are owned cooperatively. Governments at all levels regulate price, production, sales, and construction. Finally, the government intervenes most in nuclear energy. Ownership and regulation follow the same mixed pattern as in the electrical power arena. In addition, the federal government enjoys monopoly ownership of the radioactive fuel that powers the reactors and subjects the utility to detailed control in using it.

Thus there was significant variation in the public-private sector relationships in each of the fuel systems, but only coal was managed predominantly in the marketplace. Each of these systems is discussed below.

COAL

When the Arab members of the Organization of Petroleum Exporting Countries (OPEC) met in Kuwait in 1973 to declare the oil embargo, coal was supplying only approximately 18 percent of America's energy needs, as indicated in figure 3.2. In the preceding half century the proportion of the nation's energy supplied by coal had been steadily declining from a high of 80 percent at the end of World War I. Coal had been for many years the quintessential "sick man" of the United States energy industry (Davis, 1974, p. 46). It was managed by the least-developed and most unstable of the fuel policy systems.

In the decades following World War II the use of coal became concentrated in the electric-utility industry. By 1973 that industry was using 69 percent of all the coal produced in the country (U.S. General Accounting Offfice, 1977). Even with that growth, however, coal's share of the utility market declined from 52 percent to 44 percent between 1955 and 1973. Before 1973 utilities appeared to prefer oil to coal at twice the price, and clean natural gas at a lower cost clearly was a preferred option. Only the inability of the gas utilities to provide assured supplies kept natural gas from garnering an even larger portion of the electric utility market. In sum, coal was an

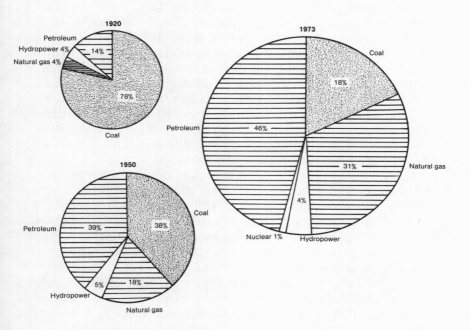

Fig. 3.2. Sources of energy supply in the United States, 1920, 1950, and 1973. From Ford Foundation Energy Policy Project, 1974a, p. 69. Copyright 1974 by the Ford Foundation.

increasingly weak competitor for a share of the growing energy market.

The coal policy system had never been capable of evolving a body of public policy that provided economic stability for the coal industry. Unlike the other fuel systems, coal policy put neither a floor under nor a ceiling on the price of coal. With a surplus supply coal production and prices fluctuated in response to short-term changes in the market.

To the extent that public policy existed for coal, until the embargo that policy was focused primarily on miner health and safety, environmental protection, and the leasing of federal coal resources. In all these concerns the effect of public policy was to inhibit rather than enhance the competitive capability of coal in the growing energy market of the United States.

In the first two decades following World War II, the primary federal involvement in coal was related to miner health and safety. Coal mining has always been one of the most hazardous activities in the entire industrial system, and coal miners have consistently suffered a higher percentage of disabling injuries than that of workers in other mining, milling, and industrial operations. The highly publicized hazards of underground coal mining, such as black lung and mine accidents, have been continuing problems.

Beginning in 1941, Congress came under significant pressure to mitigate miner health and safety problems, and laws were passed to enlarge the federal government's role in protecting miners (*Congressional Quarterly*, 1982, pp. 191–95). By the time of the embargo government programs ranged from setting health and safety standards for the industry to providing pensions to miners suffering from black lung. Over a twenty-five-year period federal legislation improved the working conditions of miners, but the fragmented nature of the industry and weak federal enforcement kept health and safety at the top of the coal policy agenda.

It was only with the enactment of the Coal Mine Health and Safety Act in 1969 that a foundation for resolution of safety issues was finally put in place. The act authorized the establishment of mandatory health and safety standards for all underground and surface coal mines, set dust and noise limits, required medical examinations and benefits for disabled miners, and provided for research programs in coal-mine health and safety. The legislation of 1969 resulted in a significant improvement in the coal industry's safety record. But these new, stricter standards did not end the controversy over mining safety. Critics of the legislation argued that it unreasonably lowered worker productivity and made coal less competitive. Supporters argued that, while the legislation was valuable, it included standards that were too lenient and urged further legislation to reduce risks in the mines (Freehling, 1973). These differing views pitted miners against operators in continuing labor-management conflicts (Hoerr, 1974).

In the early 1970s the new concerns over environmental protection added to the conflicts revolving around coal. With the passage of the Clean Air Act and its amendments of 1970 a framework was established within which the Environmental Protection Agency (EPA) was made responsible for both the establishment of national air-quality standards and specific technological requirements. Using this new legislative base, the EPA initially pressed many utilities to move away from the use of coal and toward the utilization of natural gas or oil. All this took place between 1970 and 1973.

At the same time the EPA initiated a major effort to develop cleanup technologies which it then required the utilities to use. These actions further weakened the competitiveness of coal.

The initial air-quality concern associated with coal during this early period was the emission of sulfur into the atmosphere. It was thought that one way around the sulfur problem was a general transition to the use of low-sulfur coal. The largest quantities of low-sulfur coal are in the western United States and especially in the Powder River Basin of Wyoming and Montana. Much of that coal is owned by the federal government, a circumstance that in theory might have allowed federal policy to encourage at once expanded use of coal and decreased sulfur emissions. No such action was taken, however. Rather, using the requirement in NEPA for an environmental impact statement (EIS), the Sierra Club took the secretary of the interior to court and obtained an injunction against additional federal leasing until an adequate EIS was prepared (*Congressional Quarterly*, 1981, p. 133).The inability of the coal policy system to integrate environmentalists resulted in their use of the courts to block federal coal development.

Underpinning these multiple impediments to the use of coal as a source of energy was another controversy that had been building during the late 1960s and early 1970s. That controversy revolved around the impacts of surface mining. The scarred earth resulting from surface mining in the coal fields of Appalachia and the Middle West provided a highly visible symbol of the adverse consequences of the process. Increasingly, the public became aware of such problems as scenic degradation, subsidence, spoil banks, erosion, and water-table disruption, and the coal industry saw itself threatened with requirements for reclamation that it argued would be costly and make coal even less competitive (Executive Office of the President, 1977). Unlike the issue of air pollution, however, the coal industry was able to resist the passage of comprehensive reclamation legislation until after 1973.

In sum, public policy with regard to coal contained no economic regulation and, therefore, no financial protection for the coal industry. There were, however, growing numbers of requirements focused on health and safety and environmental protection that left most observers with a pessimistic view of the potential role of coal in the nation's energy future.

From the industry's point of view, public policy for coal was uniquely negative. In only three areas was the federal government doing anything to facilitate coal development. First, the U.S. Geological Survey (USGS), as part of its general responsibilities for geologic mapping, assisted the industry by providing resource information. Second, the Department of the Interior had a limited program of research and development (R&D) support aimed at improving the efficiency of extractive technologies and processes. Third, although the EPA was viewed by the industry as a major obstacle to the development of coal, the agency did assume substantial responsibility for paying the costs of developing cleanup technologies.

Thus until the imposition of the embargo the primary focus of public policy for coal was on health and safety and environmental protection. Unlike the policies for the other four fuels, to be discussed below, the policy for coal failed to provide a stable economic environment. Coal was allowed to compete in the energy marketplace for the most part unassisted by government. The result was that coal, although the most abundant of the nation's fuel resources, was clearly the resource least capable of competing successfully.

OIL

In 1973 petroleum was supplying 46 percent of the nation's energy (figure 3.2), and, measured in absolute dollars, the unit cost of oil had steadily declined from 1948 through 1972

(O'Toole, 1976). Oil thus nicely met the standard of being cheaper with increased use, and it also had the advantage of being highly versatile because of its ease of handling, portability, and almost universal applicability.

Unlike coal, petroleum has a long history of evolving public policy to ensure adequate, low-cost supplies plus a stable profitable environment for producers. The policy arrangements begun in the 1930s were formulated to a great extent to prevent the wasteful production of oil that characterized the early days of the industry. The legal ownership of oil had been determined by the "rule of capture." Simply put, it was your oil if you pumped it out of the ground. Responding to that rule, oil producers had glutted the market, with the result that the price of petroleum had dropped to ten cents a barrel. The governments of oil-producing states and many members of the industry recognized that this pattern was dysfunctional. Not only did it waste a valuable resource, but it was economically disastrous for the industry.

The solution was the establishment of a policy system with three major components:

1. State regulatory agencies, such as the Texas Railroad Commission, allocated production levels to each well in their states. Allocations were made on the basis of a determination of the market for petroleum.

2. The Interstate Compact to Conserve Oil and Gas established in 1935 provided the legal mechanism for cooperation among the oil-producing states (Davis, 1974, p. 59). Interstate compacts are treaties among states that must be approved by the Congress. The Interstate Compact Commission established by the 1935 treaty served as the mechanism for coordinating the allocation of petroleum among producing states.

3. The Connally Hot Oil Act, passed in 1936, made it a federal crime to sell oil interstate that was produced in excess of the production allowances determined by the individual state regulatory commissions. Each state had responsibility for the enforcement of intrastate oil sales, and the federal government provided the control and management authority for interstate sales.

Two ingredients made this policy framework border on the

55

elegant. First, it allowed a floor to be put under price without ever explicitly addressing the controversial issue of pricing. Production, rather than price, was regulated, and the system had the convenient rationale of conserving oil. Second, the system acquired the authority of the federal government while at the same time providing barriers against any federal control of price that the oil-producing states and the industry did not find palatable.

Many critics of the oil industry have noted correctly that the industry generally supported the first price-regulation efforts. From that they conclude that the industry favors regulation when it is beneficial and opposes it when it is costly. Two comments are warranted in this connection. First, such behavior does not make the oil industry different from the participants in any other policy system. Second, the arrangements put in place during the 1930s effectively served the national purpose of providing ever-larger quantities of oil at ever-lower prices.

By the mid-1960s a domestic surplus of oil was no longer a problem. The regulatory organizations created to control production continued to enforce the laws, however. For example, enforcement of the Connally Act was the responsibility of the USGS. During the early 1960s the personnel of the USGS identified a growing number of wells in Texas that were producing in excess of the allowance established by the Texas Railroad Commission. In a number of instances drillers had used slant-drilling techniques to produce from reservoirs that were not under their lease property. Using information compiled by the USGS, the Justice Department initiated action against a number of these technically illegal operations, and these enforcement actions built a fire of opposition among some members of the oil industry. The opposition became so great that in 1966 the Johnson administration dismantled the component of the USGS that had enforced the Connally Act.

At the time of the embargo the mechanisms for controlling oil production were still officially in place. They were the state regulatory agencies, the Interstate Oil Compact Commission, and the regulations in the Code of Federal Regulations derived from the Connally Hot Oil Act. Eliminating the enforcement body in the USGS, however, substantially eliminated the mechanism for production control. In sum, once the need for controlling supply had ceased as a consequence of the removal of the surplus-oil problem, the oil policy system was able quietly to dismantle the production controls that had been in place for thirty years.

A second, more controversial element of policy for oil established during the days of abundant petroleum involved income redistribution. A set of tax advantages for the oil industry had been written into the federal income-tax system in the period of oil surplus. These advantages included the depletion allowance and a set of rules that allowed the industry to deduct a major portion of the costs of exploration and drilling from their taxes in the year the costs were incurred.

Under the provisions of the depletion allowance, petroleum operators were permitted to deduct 27.5 percent of their gross income from their taxable income, up to 50 percent of the taxable income. For example, if a company had a gross income of $10 million, it could deduct $2.75 million from its taxable income (Internal Revenue Code of 1954). If its net income or profit was $5.5 million, it had to pay taxes on only $2.75 million. In addition, the oil operator could deduct all the intangible costs of drilling wells and bringing them into production incurred during that year.

These tax advantages were the object of intermittent criticism over the years, since they permitted the oil industry to pay a lower rate of taxes on profits than the rate imposed on most other industries. Again, however, it must be noted that these favorable tax structures served the general purpose of producing cheap energy while allowing for predictable profits for the industry.

As might be expected, commitment to the preservation of these tax advantages was widespread in the oil policy community. But criticism of the depletion allowance grew until 1969, when President Lyndon B. Johnson proposed reducing it. After substantial debate and skirmishing the allowance was cut to 22 percent (Tax Reform Act of 1969, Title V). The depletion allowance was at that level at the time of the embargo. Meanwhile, other adjustments were made in the tax system for oil, but the industry was still allowed to deduct many of its intangible costs.

A third element in the policy for oil was put in place in 1950, and it also had income-redistribution implications.

Against a background of growing concern in the U.S. Department of State about the instability in the Middle East, Ibn Saud, the King of Saudi Arabia, demanded more money for his oil. The American oil companies involved in the Arabian American Oil Company (Aramco) sought the assistance of the United States government. World oil prices were stable; therefore, the only way for the firms to respond to Ibn Saud's demand was to increase the royalty payments. Understandably, the companies were resistant to higher payments, since they would mean lower profits.

The issue ultimately was dealt with by the United States government, which allowed American oil companies to deduct tax payments to the governments of oil-producing countries from their domestic income taxes. Under this arrangement Saudi Arabia gained additional revenues by levying taxes on firms producing within its borders, while those firms simply subtracted the tax payments from those they were making to the United States. Thus in 1950 Aramco paid American taxes in the amount of $50 million and Saudi royalties of $66 million, while in 1951 it paid only $6 million to the United States, and $100 million to Saudi Arabia (Davis, 1978, pp. 74–75). The goal of providing abundant, low-cost energy while assuring the profitability of the oil industry was maintained.

In 1959 another element in the constellation of oil policy was put in place when mandatory import restrictions were imposed on petroleum coming into the United States. Under the initial arrangements imports were limited to no more than 12.5 percent of domestic production. Responsibility for enforcing this quota system rested with the Department of the Interior (DOI). Before importing oil, the buyer must obtain a license from Interior.

Several converging factors were responsible for President Dwight D. Eisenhower's initiation of a quota system under the authority granted him in existing legislation (the Trade Agreements Extension Act of 1955). First, although domestic petroleum production increased steadily from the end of World War II until 1970, it did not grow as rapidly as consumption. The result was that imports grew from less than 1 percent of United States consumption in 1947 to 24 percent of consumption in 1970 (Stobaugh, 1979, pp. 16–18). Had it not been for import controls, the portion of the United States market supplied by imports clearly would have been much larger. Throughout much of this period imported petroleum was selling for approximately one-half the price of domestic oil.

The objectives behind the establishment of the quota system, therefore, were twofold. First, without government protection the domestic oil industry would have been severely damaged. Thus American producers were responsible for generating a great deal of pressure within the oil policy system for some kind of protectionist action. Second, following the war of 1956 between Israel and Egypt, the United States was made aware of the vulnerability of its oil imports. National-security concerns made a viable domestic oil industry an important issue and added weight to the pressures for the establishment of the quota system.

As American consumption of oil expanded, however, domestic producers were less and less readily able to meet demand. As a consequence, national policy was modified to allow a series of incremental adjustments in the quota framework. Each of these adjustments allowed the percentage of imported oil to increase vis-à-vis American production. By April, 1973, shortly before the embargo, the combination of rapidly rising consumption and now declining domestic production had brought a need for such a rapid increase in imports that President Richard M. Nixon abandoned the idea of quotas entirely (Executive Office of the President, 1973).

The protection of United States industry from cheap imports, however, was continued by levying an import fee of one cent per gallon on crude oil. This fee reduced the difference in price between domestic and imported oil sufficiently to provide continuing support for domestic firms. This was the mechanism being employed to constrain imports at the time of the embargo.

The final element in the structure of oil policy was the federal government's approach to the management of oil and gas on public lands. These lands were divided into three categories: (1) onshore lands, (2) naval petroleum reserves, and (3) the outer continental shelf (OCS).

Under the provisions of the Mineral Leasing Act of 1920 the federal government leases onshore lands for petroleum and natural-gas development under arrangements that are economically favorable to industry. Most of the leases require the

59

lease owner to pay one-eighth of the value of the production to the federal government in royalties. Most of these leases were allocated on the basis of a lottery system with a small portion allocated through competitive bids. Although the appropriateness of the financial return to the federal treasury and the income-redistribution implications for the industry were questioned by some, federal leasing was not a controversial issue in 1973.

What was heatedly debated was the increasing movement by the national government to restrict oil-and-gas development on federal lands. In response to growing environmental concerns, there was pressure to maintain many of the undeveloped western lands in their wilderness state. The petroleum industry generally was opposed to any policy that restricted oil-and-gas exploration and extraction on public lands.

In the 1920s Congress designated certain oil-producing or potentially oil-producing areas on federal lands as Naval Petroleum Reserves (NPR). The rationale behind this move was national security. During a time of national emergency or war the oil from these lands could be made available to the U.S. Navy. The state of development of these reserve lands varied. On the Elk Hills Reserve in California, for example, at the time of the embargo the Navy had carried out major drilling and development activities, and production was available immediately. At the other extreme, on Naval Petroleum Reserve Number 4 (NPR-4), a huge area in far-northwestern Alaska, only a limited exploratory drilling program had been undertaken. At the time of the embargo the Navy was funding some exploratory activities in NPR-4, but, aside from two small gas fields that supplied a naval-research facility and the city of Barrow, there was no production on these lands. Criticism of the naval reserve program focused on its inability to make oil quickly available from the undeveloped lands, but that too was not a highly charged issue in 1973.

Lands in the third category, OCS lands, were, by the early 1970s, the focus of a highly charged debate. After an intense struggle between the federal government and the coastal states, the Congress passed two pieces of legislation in 1953 that in effect designated those resources beyond three miles (or in some cases, three leagues) as belonging to the federal government. With the passage of this legislation the Department of the Interior initiated a leasing program. Development, particularly in the Gulf of Mexico, began on an accelerated basis. Over a million barrels of oil per day were being produced from the OCS at the time of the embargo (Kash et al., 1973).

Development in the Gulf of Mexico was undertaken quietly and without much debate, but there was opposition to exploration and production on the OCS off the California coast by some residents of nearby communities. Until 1969, however, the opponents of this development were fragmented. In that year the famous Santa Barbara oil spill occurred. As mentioned earlier, that event is widely believed to have been the catalyst for the mobilization of a broad cross section of the American population in support of environmental protection. Certainly the Santa Barbara accident served as a major impetus for the passage of the National Environmental Policy Act. One of the immediate effects of the spill was to bring the leasing of OCS lands to a halt pending a review of the USGS's regulatory system. At issue was whether the USGS regulations were stringent enough to guarantee environmental protection (U.S. Department of the Interior, 1980).

Following the spill and the passage of NEPA, the industry and a collection of environmental groups were pitted against each other in a series of legal actions. Some environmentalists were opposed to any oil-and-gas development on the OCS, while others took the position that the regulatory system should be much stronger and the industry much more rigorously regulated. Two opposing pressures arose in the period immediately following the spill. The first was a growing demand for federal efforts to protect fragile ecosystems, and the second was concern that domestic oil production was in a period of decline and needed to be augmented in any available way.

Today it is difficult not to view the struggle over OCS oil development as a surrogate for a much broader set of concerns that derived from the impending energy crisis, on the one hand, and the rising environmental consciousness of Americans, on the other. The OCS proved to be a particularly attractive surrogate for the energy-environment struggle. The industry believed that the OCS held the greatest potential for large new reserves of oil, while environmentalists saw in offshore petroleum resource development a direct threat to coastal ecosystems. Moreover, because the OCS was clearly and explicitly within the federal domain and therefore susceptible to direct political pressure by actors both within and outside the oil policy system, all the ingredients for protracted conflict were present.

As President Nixon became aware of the growing potential for energy problems, and as he also sought ways to divert attention from the Watergate scandal, he called for a tripling of leasing on the OCS in April, 1973. To resolve the conflict between development and environmental protection, he charged his Council on Environmental Quality with responsibility for a one-year study to formulate arrangements for the development of OCS oil and gas that would adequately deal with pollution and other externalities (see Executive Office of the President, 1973).

In the years leading up to 1973, the oil policy system demonstrated an impressive ability to use public policy to ensure stability while at the same time responding to changing conditions. But the embargo and subsequent oil scarcity demanded changes that a system established to manage surplus could not adequately provide.

NATURAL GAS

Although petroleum was the trigger for the energy crisis, the first evidence of impending trouble came from the natural-gas policy system. During the 1960s gas was the cardinal example of an abundant, cheap, clean energy source. Between 1960 and 1973 the production of natural gas in the United States rose from 13 trillion cubic feet (TCF) to 22 TCF per year. Natural gas was supplying 31 percent of the nation's energy at the time of the embargo (figure 3.2). Natural-gas policy was the best illustration of public policy designed to ensure low-cost energy.

In the mid-1960s some gas utilities began refusing to serve new customers on the grounds that they were unable to increase their gas reserves. Home builders who had been installing gas furnaces had to switch to other heat sources, such as electricity or oil. In 1970 and 1971 the Federal Power Commission (FPC) carried out studies of the gas-reserve situation and began developing schedules of priorities for supplying gas customers in anticipation of shortages (Federal Power Commission, 1970).

In the coal industry the early pressure for formulating public policy came from the miners; in the oil industry it came from the industry itself. The pressure for developing and implementing public policy for natural gas, however, came from consumers. The primary objective of natural-gas policy was, therefore, to ensure low-cost, stable supplies.

The delivery of natural gas to consumers has three phases: production, transmission, and distribution. The physical character of natural gas played a major determining role in the development of policy. Gas can be efficiently transported only through pipelines. Since it is neither economically nor practicably feasible to have competing firms delivering gas to consumers, the transportation requirements made it inevitable that gas supply would be a monopoly activity. Gas policy thus stemmed from the desire for low-cost, stable supplies and the monopolistic character of the system that supplied the gas.

Natural gas produced and consumed within a state was controlled by state government in much the same manner that intrastate electric power was controlled (discussed below). The dominant policy concerns at the time of the embargo, however, were with interstate gas. Large pipeline companies delivered interstate gas to local utilities or cities, which provided for its distribution. The FPC had control over the price that interstate pipelines paid producers, the construction of interstate pipelines, and the price charged by those pipelines for the transmission of gas (see Bupp and Schuller, 1979). Interstate transmission prices were based on a formula that took into account original costs of the pipeline, capital depreciation on that investment, and allowances for working capital. The policy system guaranteed the transmission companies an adequate profit. Despite the complexity of the pricing arrangements, there was remarkably little controversy over the formula used to establish the prices charged by transmission companies.

The FPC's control of the construction of facilities involved a process by which transmission companies obtained approval before undertaking the construction of new pipelines or the expansion of old ones. The rationale behind this approach was that such construction must be demonstrated to be in the public interest. This usually meant that there had to be sufficient demand for natural gas to pay for the cost of the facilities as well as ensure a reasonable profit. At the same time, however, the price to distributors could not be unacceptably high. Moreover, the transmission firms must be able to demonstrate that they had adequate gas to assure consumers a stable supply and adequate financial resources to pay for the construction and operating costs of any new facilities they built. While some of the FPC's decisions were controversial, the rules for resolving the disputes were widely understood and accepted.

The FPC's oversight of the price that transmission companies paid for gas at the wellhead, however, was the subject of a very heated debate at the time of the embargo. Most important to this controversy were the very substantial difficulties involved in calculating a fair return to the producers for their natural gas. Because the investments of gas producers included not just the cost of drilling, developing, and producing the gas wells but also the costs of all the dry holes or uneconomic wells resulting from their search for energy, producers consistently argued that the FPC's established gas prices were too low to make the search for new sources economically feasible.

By the early 1970s two things were evident with regard to the price the FPC allowed to be paid for gas going into the interstate gas pipeline network. First, natural gas, on a Btu basis, was much cheaper than oil. Second, interstate gas was selling for less than intrastate gas. At that time the FPC was allowing a maximum price of 50 cents per thousand cubic feet (MCF) for interstate gas. That price could rise essentially in parallel with overall inflation in the economy. In fact, the weighted average paid for gas going into the interstate network was substantially less than 50 cents per MCF, since many of the gas transmission companies had signed long-term contracts with producers at much lower rates.

FPC regulation had thus created an interesting conundrum. On the one hand, gas was a much cheaper source of energy than oil when supplied through the interstate framework. On the other hand, gas supplied to consumers within the producing states was more nearly comparable to the price of oil. Even with the addition of transmission charges gas might be cheaper on the East Coast than it was in Texas. At the same time the price paid for gas going into the interstate system was too low to permit oil-and-gas companies to search for new supplies of gas. Natural-gas producers understandably preferred either to sell their gas within the producing state or, alternatively, to cap their wells—that is, not sell the gas. As an illustration, at the time the FPC had a maximum price of $0.50 per MCF, some gas was being sold at the wellhead in Louisiana and Texas for as much as $2.24 per MCF (Davis, 1978, p. 132).

It was clear to everyone, including the FPC, that something had to be done to increase the incentives for supplying gas to the interstate market. The FPC undertook a number of incremental adjustments in the wellhead price of gas. The position of the producers was that gas should simply be deregulated at the wellhead. Consumer interests, however, argued that if that was done the price of gas would take a very sudden and very large jump, with the result that the producers would enjoy an unearned, or "windfall," profit. That is, gas that had been produced in a period when prices were lower would suddenly be made more valuable to the producing companies, and these firms would gain a substantial increase in income without having made any additional investment. Consumers supported their argument by pointing out that the nature of the pipeline system made anything resembling a free market impossible. Confronted with these arguments, the FPC found itself in a dilemma. If the price of gas was not increased, there would be no incentive for petroleum and natural-gas operators to explore for and extract new resources. If the price was increased, there would be windfall profits.

By 1973 the FPC's regulation of the wellhead price of gas was widely viewed as the cause of the growing shortage of natural gas in the interstate system. Gas curtailments by transmission companies began in 1970 and rose to about three trillion cubic feet (TCF) by 1976. In April, 1973, President Nixon went on television and asked Congress to modify the system of controls. In his speech Nixon noted that not only had price controls eliminated incentives for producers but the regulations had artificially stimulated demand for this valuable resource.

The problems with natural-gas policy in 1973 were exacerbated by the new demand for clean energy. The Environmental Protection Agency's efforts to induce electric utilities to switch

65

from coal to natural gas had increased the consumption of gas. By 1973 the commitment to a stable supply of low-cost natural gas and the rising demand for clean energy had created a rapidly accelerating pattern of gas use that experts agreed would soon lead to a crisis. Since the demand for natural gas would likely be highest in the coldest part of the winter, it appeared that federal regulations were going to be a major contributor to either shutting down a number of industries or leaving many Americans cold.

Some observers of the oncoming crisis saw the import of natural gas from Canada and of liquefied natural gas (LNG) from Algeria as at least a partial solution. The FPC's commitment to maintaining low-cost gas, however, complicated negotiations in both of these areas. Canadians were not willing to permit a United States federal regulatory agency to determine the sales price of their natural gas, and it was a given that the price of importing LNG would be much higher than the regulated price of domestic gas.

Thus by 1973 national policy for natural gas was the object of a major controversy. The low-cost, clean-energy goals of that policy had taken precedence over the adequate-supply goal. By the time of the embargo there was not really a question whether policy modifications in the system would be necessary; rather the question was when and what kinds of modifications would be made.

It must be emphasized that policy for natural gas had been a major factor in the rapid growth of gas as an energy source. That policy had ensured abundant, low-cost supplies while guaranteeing profits to the industry. Like the oil policy system, the gas policy system was capable of adjusting to changing circumstances in a context of surplus. Scarcity, however, demanded changes that the policy system was not capable of making.

ELECTRICITY

Like natural gas, electricity is a natural monopoly. That is, the benefits of having multiple electrical systems competing for the same market had not been judged to be equal to the benefits of having a single, regulated system. By 1973 a workable electricity policy had been achieved, though not without struggle and debate.

In the years preceding the embargo the consumption of electric power in the United States grew at a rate of about 7 percent a year, almost twice the growth rate of overall energy consumption. Like natural gas, electricity was a highly valued source of energy because of its flexibility in meeting a range of end-use needs. Jerrold Krenz (1980, p. 26) has characterized this valued source:

Electricity can be considered a form of highly processed energy, and as such it can be used for many applications in which fuels are less suited. For example, illumination can readily be produced by an electric incandescent or fluorescent bulb and with considerably less fuel used by the power plant than by an individual oil lantern or a gas jet. Another example is an electric motor. Even considering the 30% efficiency of an electric power system, a 90% efficient electric motor (as is common for large units) has a higher overall efficiency based on fuel used by the power plant than a gasoline or diesel engine. (In addition, an electric motor is frequently more convenient than an internal-combustion engine.)

The electric power system in the United States involved multiple sets of public regulatory agencies, multiple sets of generating organizations, and a complex mix of transportation and delivery bodies. Among all these groups there was, in the early 1970s, a consensus on two things. First, they agreed that electric power was a right—that consumers of electric energy had the right to all of it they wanted. From historical experience this meant that the electric power system had to bring on line sufficient generating and distribution capacity every year to take care of the 7 percent annual growth rate. Second, both the regulated and the regulators subscribed to the view that to meet this demand there were substantial advantages in opting for economies of scale in electric facilities. Thus the electric power policy system supported ever-larger generation and transmission facilities, basing that policy on a long record demonstrating that huge facilities led to lower unit costs.

Like natural-gas operations, electricity operations can be divided into three segments: generation, transmission, and distribution. Unlike natural-gas operations, however, electricity operations do not follow a fixed pattern. In some locations the same company is responsible for generation, transmission,

67

and distribution of electricity both within the state and across state lines. In other areas generation is the province of one body, transmission another, and distribution still another, as with natural gas. Also, unlike coal, oil, or natural-gas operations, government itself—federal, state, and local—is involved in electric power, operating as generator, transmitter, and distributor. As an illustration, privately owned firms generate between 75 and 80 percent of the nation's electricity, and the federal government generates in excess of 10 percent. State and local governments and rural cooperatives generate the remaining 10 to 15 percent (Davis, 1978, p. 171).

During the period of development of electric power, the objectives sought by public policy were multiple, and the history of electric energy is a unique and fascinating tale. All these early struggles revolved in one way or another around the best way to develop a system that would ensure abundant, low-cost electricity. By the early 1970s the rules of the game were widely accepted and the source of little friction. Adoption of the new goal of clean energy, however, brought major conflict.

Abundant, cheap electricity was pursued under a set of public-policy rules that were rather simple in concept though frequently complex in implementation. Those rules were that generation, transmission, and distribution organizations should be allowed to make a fixed-percentage profit on their capital investments and operating costs. Such rules applied, in their own unique ways, to all the various state regulatory organizations as well as the FPC, which represented the federal government in this policy system. The federally owned generating, transportation, and delivery organizations, such as the TVA, provided a yardstick by which regulatory agencies could evaluate the information provided by electric utilities. In few areas have government regulatory bodies had more detailed access to the activities of an industry than in electric power. The primary struggles between government oversight organizations and the electric utilities centered on the appropriate way to value the capital equipment of the utilities. Negotiations between regulators and utilities, therefore, were infinitely complex and technical, but they did not involve basic differences over what the regulatory strategy should be.

The technology of the electric power industry, specifically the large facilities that were established because of their economies of scale, gave regulatory organizations and the utility industry a common interest. The utilities could make more money by investing ever-larger amounts in capital equipment. The larger the investment, the greater the profit, when the profit was a fixed percentage of that investment. At the same time those large investments led to lower and lower unit costs.

On the consumption side of the equation, this situation made it advantageous, in the eyes of both the regulators and the regulated, to supply cheaper energy to the largest users. The electric power industry commonly supplied electricity to large industrial consumers on a declining-rate basis: the unit cost of electricity decreased as the quantity used increased. There appeared to be a unique natural benefit both to the consumer and to the producer that was inherent in the technology.

Certain problems developed very early with regard to public policy for electric power. One of the first problems was how to prevent fraud and excess profits in the industry. In the early days abuses in the electric power industry resulted from multiple holding companies. Under a complex set of arrangements an operating firm would be owned by a holding company, which might be owned by still another holding company. Many of these holding companies provided no services themselves but nonetheless extracted large fees. The fees in turn were fed into the rates of the operating company, to the ultimate disadvantage of the consumer. In 1935, in response to this situation, Congress passed the Public Utilities Holding Company Act, which prohibited multiple layers of firms but also delegated, initially to the Securities and Exchange Commission and later, by example, to the various state regulatory agencies, close involvement in the financial activities of the utilities.

The utility industry did not develop a pattern of retaining earnings and using those earnings to invest in new capital equipment. Rather, the profits of the utility companies were distributed to shareholders. Money for new capital investments was raised either by selling stock or by borrowing, usually through long-term bond issues. With interest rates remaining steady and relatively low for many years, and with the guaranteed profit of the utility companies resulting from their regulation as natural monopolies, it was easy to raise capital. The pattern of raising capital funds by borrowing or selling stock also made regulation easier. The manner in which the electric utilities generated their capital was easy until 1973, but in the wake of the embargo it became a very serious problem.

While electric utilities generally were organized on a local, state, or regional level, the large size and cost of generation facilities plus the efficiencies that resulted from transferring electricity from one company to another led the FPC to encourage the development of large regional electrical grid systems. Such systems, it was believed, would assure increased reliability of service by making it possible to transfer electricity from one utility to another. If a power plant belonging to one utility failed—in the terms of the industry, "went down"—it could buy power from an adjoining system. And so it went across the country. Only after the widespread blackout of the northeastern United States and southeastern Canada in 1966 were questions raised about the effectiveness of the electrical grid system. Even after the power failure of 1966 the primary criticism of the FPC was that its policy, a truly unique action for a regulatory agency, had pushed the production, transport, and delivery system ahead of the industry's technology.

All in all, however, the system worked, as is evidenced by the continuing assumption that access to electricity was a right. By the late 1960s the future of the electric power industry seemed increasingly to rest on energy from nuclear power and coal resources. Unfortunately, these were the two sources most vigorously opposed by the environmental movement and most in conflict with the emerging policy goal of assuring clean supplies of energy. Because the construction of large coal and nuclear electric power plants takes many years and involves huge inputs of capital and expertise, these facilities provided exactly the kind of focus around which it was easy to mobilize opposition. Thus the electric power system found itself in the unique position of offering the cleanest form of energy at the end-use point—electricity—while at the same time relying on production processes that were among the most controversial.

At the time of the embargo the electric power industry was being whipsawed by these conflicting demands. Between 1970 and 1973 the EPA put pressure on the industry to move away from coal. As we shall see, however, when the embargo was imposed, the government reversed direction and encouraged the industry to move rapidly back to coal sources for the generation of electricity. Both the technology of electric power and the policy in place in 1973 put the industry in a very difficult situation. The energy crisis exacerbated these difficulties.

Thus, like the oil and gas policy systems, the electric power system worked well until the period immediately before the embargo. It supplied low-cost, abundant electricity and provided assured profits. It was, however, ill-equipped to deal with perceptions of scarcity that made both its market and its sources of primary energy uncertain.

NUCLEAR POWER

Of all the sources of energy, nuclear power is the most explicitly a gift, or, alternatively, a curse, of the federal government. Nuclear energy had its beginnings in the Manhattan Project of World War II, which produced the atomic bombs that were dropped on Japan. The project was managed by the U.S. Army Corps of Engineers under arrangements aimed at ensuring tight military security. Although control of nuclear energy was transferred to a civilian organization, the Atomic Energy Commission (AEC), in 1946, the pervasiveness of national-security concerns has been a factor in the subsequent development of nuclear power. The national-security implications of all aspects of nuclear power, added to the development of the technology by the federal government, ensured that government would remain the predominant and pervasive force in the development of the resource.

No other energy option has been so surrounded by mystery and uncertainty. Few laymen have even an elementary understanding of the processes whereby energy is released in nuclear reactions and controlled and channeled into electricity. Security concerns have heightened the uncertainty.

As one looks back at thirty years of development of nuclear energy, it is hard to conclude other than that the mystery surrounding things nuclear has polarized perceptions about the potential costs, risks, and benefits associated with this technological alternative. Uncertainty has magnified reactions to nuclear power across the board. As indicated earlier in this chapter, at one time it was possible to believe that nuclear

71

energy would be the nation's energy savior. This arcane activity could, it was claimed, produce electricity so inexpensive that Americans could achieve the ultimate in their goal of unlimited, cheap power. At the other extreme it has been possible to believe that nuclear power is an evil genie let out of the bottle—a genie that would bring civilization to an end either through a cataclysmic military exchange or a more insidious poisoning of the environment with radioactivity, the ultimate hazard. In no area of energy policy have there been so many "facts" and "explanations" that have contributed so little to the development of a public consensus.

The policy in place for nuclear power at the time of the embargo was predominantly the creation of a community of institutions and individuals who saw nuclear energy as a shining hope. The policy for nuclear power was constructed and implemented by a self-contained, uniquely protected policy community. This community moved from nuclear weapons to civilian nuclear energy through three stages: the development of weapons technologies, the development of nuclear-powered submarines, and the innovation of nuclear-powered electric generation plants. Each of these transitions involved policy debates and struggles that are unequaled in complexity and uncertainty.

The U.S. Navy became committed to the development of nuclear-powered submarines in the 1950s. The goal was a nuclear power plant that could be substituted for diesel power. As a part of this development generation of electricity by nuclear power was first demonstrated in 1951. Basing its technology on the reactor developed for submarines, the AEC moved to the task of producing a commercial electric facility that would generate power from a nuclear reaction. As it turned out, the most successful of the commercial nuclear-reactor designs, one produced by Westinghouse, basically was an enlarged version of the submarine power plant.

Civilian nuclear energy policy had two dominant objectives. One was the rapid development and use of nuclear energy to provide commercial electricity. Nearly all the power produced by nuclear reactors is distributed and consumed in this form. The second objective was to ensure that the widespread use of commercial nuclear energy did not provide to other nations a military nuclear capability. The attainment of these two goals required that the federal government have a continuing, detailed involvement in the development and use of nuclear power.

Discussions of the nuclear power system revolve around the nuclear "fuel cycle." That set of activities begins, in the simplest form, with the mining of uranium and continues through several processing and refining steps to the generation of electricity and the reprocessing of used nuclear fuel, one part of the reprocessed fuel going back to the electric power generation point and another part going to some kind of permanent storage. A basic outline of the fuel cycle is provided in figure 3.3.

Before the oil embargo the federal government assumed responsibility—in fact, demanded the right to be overall manager and regulator—of the entire fuel cycle. As it became possible for entities in the private sector to make a profit carrying out the functions involved in one or another of the fuel-cycle stages, those stages were turned over to private firms. When certain steps of the fuel cycle did not appear profitable, the federal government assumed responsibility for paying for the research, development, and demonstration costs. By 1973, the responsibility for building the first-generation nuclear electric plants in this country had been taken over by several corporations in the private sector. These companies designed and built the plants and delivered them to the electric utilities. As far as the reactor portion of the fuel cycle was concerned, once these private interests had become involved, the primary attention of the federal government was focused on the development of second- and third-generation nuclear power facilities. At the time of the oil embargo the federal government was paying a major portion of the costs of a breeder-reactor demonstration plant at Clinch River, Tennessee. The plant was partly funded by electric utilities and other corporations, but the liability of these private actors was fixed, and the federal government absorbed all the costs and associated risks beyond that fixed dollar amount.

The pattern followed in developing the breeder was essentially the same pattern that had been followed in each stage of the first-generation nuclear fuel cycle. The federal govern-

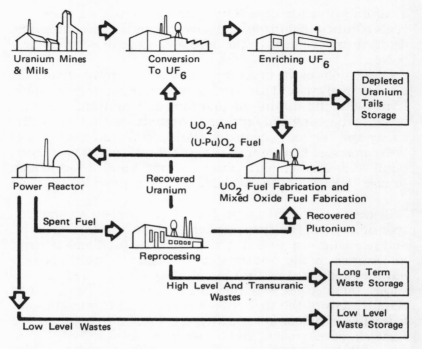

Fig. 3.3. The nuclear fuel cycle. From *Nuclear Power Issues and Choices: The Report of the Nuclear Energy Policy Study Group,* 1977, p. 403. Copyright 1977, the Ford Foundation.

ment initiated the efforts that led to the development of a uranium mining industry by guaranteeing a market for the resource that was mined. It also paid for the development of a number of the steps in turning uranium ore into a usable fuel. The most important of those steps, the enrichment stage, was carried out at federally owned facilities. Similarly, the federal government provided financial support for the development of reprocessing technology. The first reprocessing plant was paid for by the government and taken over by the private sector when it appeared to be profitable. When problems developed with the plant, the federal government provided the impetus for the development of more advanced reprocessing technology.

The policy system gave the federal government control of nuclear fuel. Through this management oversight the government tried to ensure that enriched uranium or plutonium would not become available for the development of weapons overseas (see National Research Council, 1979). In fact, the AEC undertook an effort to monitor nuclear fuel around the world.

Even after the responsibility for the construction and operation of the various fuel-cycle activities passed to the private sector, the AEC established the standards and maintained the inspection system. For example, nuclear power reactors had to be constructed and operated under criteria set by the AEC, and agency inspectors continuously reviewed both the structural and the operational aspects of nuclear facilities. As we shall see, much of the criticism of the AEC's approach to nuclear power policy came from those who felt that the agency sacrificed safety for the promotion of nuclear energy resource development. In an effort to respond to these concerns, the AEC's standards and guidelines became increasingly more stringent.

Over time the licensing of nuclear electric generating plants became the focus of opponents of nuclear power. The AEC's review process provided several opportunities for outsiders to intervene, express their concerns, and make recommendations. Eventually interveners learned how to manipulate this process to delay plant construction by demanding additional studies and arguing for modifications in the design of nuclear facilities. Over the years the increasing use of this tactic led to a marked slowdown in the construction of nuclear plants. Utilities found that the lengthening of lead times in this fashion increased the cost of building nuclear facilities, and they were especially troubled by delays in the late stages of construction, when very large capital outlays made extended scheduling an economic nightmare.

As early as 1957 a major impediment to the rapid development of nuclear power developed. The risks of a reactor accident were perceived by enough people that the utilities became deeply concerned about possible liabilities. So little was known about the nuclear power industry that no insurance could be obtained in the private sector for the potentially unlimited liability that might result from a nuclear accident. As a consequence Congress passed the Price-Anderson Act to provide government indemnity to support private firms operating nuclear reactors.

By 1973 federal nuclear power policy addressed four major needs: (1) it guaranteed the government subsidy necessary to carry out the R&D in the fuel cycle until technologies could be demonstrated to be commercially attractive for companies in the private sector; (2) it limited the liability of utilities resulting from accidents in nuclear power plants; (3) through federal ownership it protected against proliferation of enriched nuclear fuels that could be used for weapons; and (4) it sought to protect the health and safety of Americans by establishing strict design, construction, and operation standards for nuclear energy facilities.

These four elements of government policy resulted in what appeared to be a commercial success with nuclear power. The movement of electric utilities to nuclear generating plants accelerated at an ever-greater rate in the early 1970s. It was expected that by the end of the twentieth century a very substantial portion of the electricity generated in the United States would be provided by nuclear facilities.

The initial impression of commercial success, however, was beginning to be severely challenged by the time of the embargo. There were multiple reasons for the change, but perhaps the key ingredient was a large gap in the government system of supports and subsidies for nuclear energy: the policy system had not formulated and implemented a way to handle nuclear wastes. In what now seems to be a rather shortsighted view, it was believed that nuclear power would develop rapidly and achieve such success that it would be possible to leave the waste-disposal problem to the utility industry. During this initial euphoric period the difficulties surrounding the permanent disposal of nuclear wastes were postponed indefinitely. By the early 1970s the wastes were beginning to accumulate, and it was widely recognized that answers to the question of permanent disposal were not going to be easily demonstrated or accepted.

Nuclear power policy had been formulated largely by those who were optimistic about the set of technologies that together comprise the fuel cycle. The complexity of these technologies and the screen of security that surrounded nuclear decisions

76

made it possible for the optimists to develop a body of policies for nuclear energy that appeared at first glance to be adequate. By the beginning of the decade of the 1970s, however, this set of policies no longer appeared adequate. Opposition to nuclear power was growing. Nonetheless, up to 1973, nuclear power was seen as a success story. Measured by the goals of abundance, cheapness, and, in the view of many, cleanness, nuclear power policy was meeting the nation's needs.

Clearly the most striking conclusion to be drawn from this review of the policies developed for the five fuels is that successful delivery of cheap, abundant energy relied heavily on public policy. Stated simply, the American experience with energy up to 1973 involved a pervasive use of public policy to establish economic boundaries for the fuel systems. More precisely, public policy was developed to ensure that the producers of the fuels, with the exception of coal producers, had dependable and acceptable levels of profit.

Given abundant fuels, all the policy systems except coal evolved arrangements that delivered abundant supplies of cheap energy while assuring their producers acceptable, dependable profits. Alternatively, it must be emphasized that the United States energy system had no experience with a truly free market system. Only coal operated without the protection of public policy. The history of the coal industry must raise serious questions about the lack of a body of public policy that can provide economic protection and stability.

It has been shown that each of the fuel policy systems was under stress by the time of the oil embargo. The Arab action focused attention on these stresses. The inadequacy of the fuel policy systems was quickly and widely perceived. In the following chapters we shall investigate the nation's efforts to respond to these inadequacies.

PART TWO

Energy Policy Developments, 1973–1980

CHAPTER 4

A Can of Worms: The Energy Policy Community After 1973

There are basic divisions in this society regarding energy. They involve disagreement about the kind of future people want, about the prospects for attaining different futures, and about the distribution of the benefits, costs, and risks from acquiring and using energy. The nature of these disagreements is much broader than those encompassed specifically within the energy sector, but the events which forced energy decisions to the fore have focused these conflicts on energy matters. Energy has become the testing ground for conflict over broader social choices. —Sam H. Schurr et al., *Energy in America's Future*, 1979

It has been seen that the fuel policy systems that had provided stability in managing the nation's energy following World War II were experiencing stresses before the oil embargo of 1973. Stress came from changing physical circumstances as well as adoption of new goals and objectives. In petroleum and natural gas stress came largely from physical change. New supplies of these fuels were not being discovered at a rate sufficient to replace production. In the coal, electric, and nuclear power industries the growing environmentalism was hampering rapid development. Until the embargo, however, each of the fuel systems was seeking to make necessary adjustments within its own bounds.

As is the norm in American policy, the first response to rapid social, economic, and technological change was to try to fit that change into existing organizations and patterns of behavior (see Grafton, 1975). The central message of the embargo, however, was that incremental modification in the existing institutions would not do. Energy would have to be managed as a system. Although from hindsight the need to manage energy seems clear, it was by no means accepted by many of the traditional participants in the fuel policy systems. Furthermore, even those actors and interests who saw the need to manage energy more comprehensively disagreed about what energy meant as an organizational concept or how it should be managed.

One of the most striking aspects of events following the embargo was how quickly the fuel policy systems' management of energy became disarrayed. Looking back, we can see that fragmentation should have been expected. Stable systems, as described in chapter 2, are extremely informal and in many ways very fragile constructs. They depend on the maintenance of consensus in a host of areas, including agreement on the activities that should be included in the policy sector and their location in the subsectors (available, potential, or theoretically possible), the parties-at-interest that have a legitimate right to participate in policymaking, and conformity to a set of decision rules. The embargo and its aftermath shattered consensus in each of these areas.

With the disintegration of consensus on the goals of energy policy, the sources of energy, the instruments of policy, and the legitimate participants in the policymaking community, energy management immediately moved to the executive-congressional agenda. What the executive and legislative decision makers faced following the embargo was a large, complex, and conflicting range of actors seeking to influence the energy policy. The many advocates who focused their attention and actions on the president and Congress shared only two things in common. First was the pervasiveness of the disagreements among the various energy interests. Not only was there no consensus regarding the factors leading to the energy crisis ("explanations" of the events of the early 1970s ranged from projections of the physical exhaustion of energy resources to allegations of conspiracies by the oil companies, the federal government, and the petroleum producing countries), but there was no agreement about even the fundamental facts upon which a future energy policy might be based.

Hans Landsberg (1980, pp. 76–78) has argued that among the factors that made consensus building in the energy system extremely difficult, the following were major obstacles:

1. United States resources have been abundant. Thus increases in cost and decreases in supply of energy are fiercely resisted.

2. Energy producers have long been the targets of popular hostility. "Big oil" conjures up the robber barons of the early Standard Oil days, coal has been branded with the antilabor stamp, and utilities have been the object of public suspicion.

3. The rise of environmental concern merely reinforced anti-energy attitudes. Pollution was intimately linked to energy production, transportation, conversion, and consumption activities.

4. Energy producers are typically big. The revival of anti-bigness sentiment found a natural target in much of the energy business.

5. The "limits-to-growth" syndrome reawakened concern about rapid depletion of the world's energy resources. The rapid post–World War II recovery of Western Europe and Japan, coupled with escalating energy demands by the developing countries, put energy resource depletion high on the list of public concerns.

6. Antigrowth sentiment increased as population and income growth worsened conditions in congested areas and encroached upon parks, wilderness areas, and other former havens of solitude. Energy, as the major engine of growth, became the proxy for growth itself to some individuals and groups.

7. Technology became a target for criticism. Energy, because it powered most modern technology, became a stand-in for antitechnology sentiment.

8. Energy-resource development facilitated the increased power of the federal government and led to confrontations between states. Energy considerations gave rise to new regional lineups.

9. Decades of living with low-cost, abundant, reliable supplies of energy had given consumers a vested interest in their perpetuation. Energy prices were singled out as the cause of general inflation.

10. A large share of the energy used by the major oil-consuming nations is imported from the Middle East. Fear of supply

interruptions was linked with highly emotional and complex foreign-policy and defense concerns.

Given the above factors, debate immediately after the embargo was perhaps the most visceral in the nation's entire energy history. Years of apathy and drift suddenly gave way to near panic and scapegoat politics. Such an environment encouraged polemics, and the results were not long in coming. As only one example of this form of dysfunctional debate, consider the following by William Riker, a political scientist writing in *No Time to Confuse*, published in 1975 as a response to the Ford Foundation's landmark *A Time to Choose*. Countering the foundation study's effort to raise energy conservation to a high-priority policy goal, Riker (1975, p. 155) argued:

This kind of ideology is used to justify the existence of a planned society which works, if it works at all, only in the context of a police state. Much of the ordinary workaday life would have to be made illegal to force society to make fuel conservation its highest priority. The electrician who wires a house for comfort rather than conservation, the salesman who speeds on his route by automobile rather than available public transit, the mayor who delays building a sewage plant because he cannot find money for it—all these will be transformed into commercial criminals. We could make "energy conservation a matter of highest national priority," as the report urges Congress to do; but the kind of life for which we conserved it would not be very attractive.

The charges and countercharges that filled the air in the post-1973 years are a clear demonstration that energy had become the irreducible requirement of our technological society.

This chapter presents a broad description of the range of actors and interests with which the president and Congress had to deal following the embargo. The nation's inability to respond rapidly to the issues presented by the events of the early 1970s resulted from two sets of factors. First was the disintegration of the relatively orderly policy communities that had made decisions about the five fuels. After 1973 there was a veritable explosion in the number and diversity of parties-at-interest seeking to influence national policymaking for energy. With this explosion the era of continuity and ongoing, stable participation by a few vested interests came to an end. The dynamics of this rapid proliferation of energy actors are the focus of this chapter. Second (investigated in the next chapters) was the expansion of the range of energy resources that were offered as answers to the energy crisis. Contributing to the fragmentation of the traditional fuel policy communities was the high degree of uncertainty about which technologies were available, potential, or merely theoretically possible.

The major dilemma the Congress and the president faced in trying to come to grips with the upsurge of energy actors after 1973 was not that they were motivated any differently from the traditional participants. In fact, the drives of performance, impact, and power were still operable—indeed, were, if anything, even more potent as energy became a high priority item. What was different was that new combinations of these motives were at work in many of the emerging participants. These combinations of motives made the actions of new interests less predictable than those of the old actors and created a source of instability. In the sections that follow, we classify these actors according to their major motives and illustrate how the motives became more complex and uncertain over time. It should be emphasized that what evolved in the years immediately following the embargo was a highly complex set of participants that bore little resemblance to the stable communities of the past.

This chapter is divided into four sections. The first section outlines the characteristics of the traditional fuel policy participants, to provide a better understanding of the magnitude of change that was introduced by the events of the early 1970s. The second section examines the new actors, motivated primarily by performance concerns, particularly the advocates of new energy technologies, and the influential energy modelers. The third section assesses the role of new impact-motivated actors—those driven to participate by short-term economic costs, risks, and benefits and those involved in energy affairs primarily because of concerns about environmental and other second-order or unplanned consequences of energy decisions. The fourth section delineates the emerging participants whose basic motivation was power. In the last category were three new kinds of players: new federal and state agency participants,

85

actors concerned with foreign policy, and those who saw energy as a means by which basic sociopolitical system changes could be made.

TRADITIONAL FUEL POLICY SYSTEM PARTICIPANTS

Although the disorder that followed the oil embargo graphically demonstrated the inadequacy of the fuel policy systems to deal with the facts of energy and fragmented their capacity to develop policy by consensus, these communities continued to play an important role as advocates to the president and Congress. In general, after 1973 the five fuel communities sought either to protect their positions or, alternatively, to expand their policy advantages. The advocacy role of these communities had a common theme: the way to deal with the energy crisis was to increase domestic production of the traditional fuels, and the major obstacle to such a strategy was excessive public-sector regulation of private-sector activities (whether of a pricing, public-land access, or environmental, health, or safety character). The traditional communities generally held the view that the energy crisis could have been avoided simply by greater reliance on the free market. It must be noted, however, that this view represented more than just a philosophy—the faith in a "producer perspective" also was a fundamental source of previous consensus and stability in these traditional communities. As Leon Lindberg (1977, p. 367) has argued:

Energy [fuel] policy systems are very resistant to change, in spite of abundant information suggesting that existing policies are inadequate or counterproductive. They uniformly resist forces for the consideration of alternative technologies or other economic development options. The "supply orientation" combined with a pervasive faith in technology produce organizational routines and "selective misperceptions of uncertainty." Established relationships with "producer groups" are powerful forces for inertia. The tradition of closed incremental and technocratic decision-making obscures broad policy. The use of mathematical models emphasizes the role of the established expert and disqualifies the outsider. The preferred practice of regulatory agencies is case-by-case, rooted in precedent and favoring the proven over the novel.

In this context it is important to emphasize that the systems did not resist all change. Stable policy systems are successful only to the degree that they demonstrate a capability for selectively incorporating change—an ability to demonstrate "persistence with and through change." Absolute stability is neither possible nor desirable. What the traditional fuel policy systems did, following the embargo, was seek to respond to a changed set of facts about energy while at the same time ensuring that the patterns of policymaking were "kept within the bounds of recognizability" (Young, 1968, p. 6). In sum, the fuel policy communities were capable of changing as long as the focus of action remained a specific fuel. Unfortunately, many of the new participants in energy policy after 1973 had no attachment to the old fuel systems, and, in fact, many of them, as will be seen, were advocates of changes so far-reaching that the destruction of the old systems was a foregone conclusion. Many of the new participants simply refused to play within the "bounds of recognizability."

It should be recalled that the first serious challenges to the stability of the traditional communities and their systems were already taking place before the embargo. The environmental and consumer movements of the late 1960s and early 1970s posed the first threats to the relatively narrow group of actors, mostly individuals in industry and government, making up these communities. But despite these challenges the fuel communities of the mid-1970s still closely resembled those of the immediate postwar period. Figure 4.1 provides an outline of the major participants in each of the five fuel communities in the mid-1970s. The linkage to the previous two decades is obvious. What occurred in the next few years, however, was that a wider range of actors and interests began seeking to intervene in fuel policies.

The essential point is that, despite change, the fuel policy communities did not disperse. They continued to function following the embargo, and although they were no longer capable of developing policy for specific fuels by consensus, they became very powerful advocates in the conflict ridden presidential-congressional decision-making process that dominated energy policy after 1973. Although their influence at the highest levels of the federal government ebbed and flowed over the

	Coal	Oil	Electricity	Natural Gas	Nuclear Fuel
Industry					
Associations Ownership	NCA Oil firms own much production	API, NPC Vertical horizontal, integration of majors	EEI, EPRI Both public- and investor-owned firms	AGA, API Close ties to oil producers, but also transmission, distribution firms	None Public monopoly until recently
Politics	Poorest energy industry, low profile	Richest energy industry—heavy lobbying and campaign aid	Limited, but increasing lobbying	Complexity limits community of interest	Dominated by scientists, atomic energy experts
Labor					
Membership	UMW, 115,000	OCAW, 200,000	No labor union that is indigenous to electricity	Same as oil industry	Same as oil, gas industries
Politics	Strongest energy union—conflict with IUOE	Weak history of unionization			
Government					
Public–private interface	Least subject to federal control—history of state control	Heavily regulated at federal, state levels	Major state role	Major state role	Most subject to federal control
Regulatory agencies	Interior Department (MESA)	FEA, DOI	FPC	FPC	NRC

Fig. 4.1. Characteristics of the participants in the fuel policy communities.

years between 1973 and 1980, the traditional fuel communities were perhaps the most consistent and certainly among the most effective advocates in the energy arena.

THE NEW PARTICIPANTS

Performance-Motivated Participants

One of the basic ongoing motives for involvement in energy policy has been performance. In chapter 2 it was noted that the performance drive has to do with designing scientific and technological policies so that they solve substantive problems. It was also observed that scientific and engineering expertise historically has provided most of the ideas around which stable policy systems have oriented their problem-solving activities. After 1973 two new sets of actors driven by the performance motivation came onto the stage that had formerly been occupied almost exclusively by scientific-technological experts from the five fuel systems. These new actors were the advocates of alternative energy technologies and energy system modelers. Actually, neither of these groups was new, but they had previously operated at the periphery of the fuel policy communities, kept away from the key centers of power by the domination of the conventional oil, gas, nuclear power, coal, and electricity experts.

Two closely related sets of factors explain the emergence as significant participants of the new performance-motivated actors. First, after the embargo it was increasingly obvious that the traditional fuels could no longer meet Americans' desire for energy that was abundant and secure (to say nothing of clean and cheap). New, nonconventional resources would have to be considered; thus the arrival of experts in alternative energy sources. Second, the heightened uncertainty about future sources demanded greater attention to forecasting supply-and-demand requirements and capabilities over longer time frames. Some anticipatory, rather than purely reactive, capability was needed at the highest level of government. Thus the expanded role for energy modelers.

Advocates of New Technologies

Immediately following the embargo a broad range of experts

knowledgeable about new or alternative sources of energy rose to lobby their preferred choices with the president and Congress. These were the advocates of new sources of energy (oil shale, tar sands, organic wastes, geothermal, solar, and conservation) and of finding alternative ways to utilize conventional fuels (for example, producing synthetic liquids or gases from coal). This book has previously argued that experts have a high degree of authority within stable policy systems. Similarly, the struggle over energy policy at the executive congressional level gave experts a prominent role. From the perspective of hindsight it may be that the policy struggle between 1973 and 1980 was too permeable to expert advice and analysis. What faced the president and Congress following 1973 was a situation in which conflicting expertise seemed to dominate nearly every issue. The lack of agreement among energy experts goes a long way toward explaining the inertia that characterized much of early postembargo decision making.

Most important to this conflict between experts, as will be explicated in some detail in the following chapters, was that the major newly proposed energy resource options were fraught with uncertainty. Experts involved with any new technology or physical activity, however, must be optimists. One who devotes many years or a lifetime to the development of a particular technology or process has a tendency to believe that the technology or process can be made to work and, in fact, that it can be made to perform even better than other options. The scientific-engineering community recognizes the need for professional optimism and therefore tends to discount it. On the other hand, nonexperts tend either not to discount the optimism or to assume that it is motivated less by admirable motives such as technical professionalism and more by vested interests, such as the desire for wealth or power.

In the immediate postembargo period the expert promoters of alternative sources of energy reflected such optimism in their characterizations of the potential contributions that could be made by new energy resources in the short to mid-term. If one characterizes an energy resource positively, that optimism must be based on some comparative measure. For example, the promoters of shale oil consistently compared its costs favorably vis-à-vis the expected costs of crude oil. Similarly, promoters of the breeder reactor produced analyses that showed it to be economically advantageous when compared to coal. And so it went with solar power, energy conservation, organic wastes, and so forth. In the promotion of the new technologies, then, the experts inevitably cast doubt on the viability of other, competing options.

When such comparisons occur within stable policy communities that evolve decisions by consensus, the discounting of optimism is ordered and effectively managed. What happened in the wake of the embargo, however, was that all the comparisons among energy resources were made in a conflict-ridden setting, where they received major publicity. Two events of this period cast doubts on the validity (to say nothing of the credibility) of the technical analyses offered by alternative energy resource experts. On the one hand, it was common for actors driven by motives other than technological performance— by impact or power concerns—to organize around a particular emerging technology in either devoted support or determined opposition. For example, environmentalists could be expected to support solar energy and conservation options and to use the most optimistic projections of the experts in each of these areas as if they were demonstrated available processes or technologies. On the other hand, supporters of nuclear power often tended to denigrate the potential of solar and conservation choices because they perceived them as instruments used by antinuclear forces to impede nuclear development. Whether or not experts chose to behave in an overtly political fashion or not became irrelevant. The line between expertise and concerns other than scientific and technical performance had been blurred.

Finally, the authority and credibility of experts were eroded in part because of the rapid increase in federally supported energy R&D activities. Critics of the new energy technologies sought to raise questions about reports of optimistic findings on emerging technological alternatives by suggesting that such reports were motivated by the desire to gain the next federal grant or contract. Similarly, new bureaucracies grew up in federal agencies to oversee the disbursement and use of these monies, and, once such an organization was in place, it was inevitable that suspicions would grow about whether it was presenting the most accurate picture of the new technologies. A continuing concern in the postembargo period was that the optimism about the performance of alternative energy processes and techniques might be flowing from the need of these bureaucracies to justify and sustain previous decisions.

In sum, in the years after 1973 energy expertise came to be viewed very differently from the view held of it when there were stable policy communities. Under conditions of stability experts had gained substantial power from the widespread perception that their only concern was performance. As stability eroded in the aftermath of the embargo, it was no longer possible to isolate "the facts" of technological performance quite so easily, and scientific and technical information became highly suspect. Both the competence and the reputation for honesty of the expert community became an issue in the debates and the struggles over what national energy policy should be. Soon both the president and the Congress were faced with the dilemma of trying to choose among competing and conflicting energy experts. Under these circumstances in the mid-1970s scientific and technological information became more of a contributor to conflict than a basis for conflict resolution and policy consensus.

Energy Modelers

A second group of energy experts quickly moved onto the political scene after 1973. These were the modelers of the United States and world energy system. Because attention had been predominantly focused on fuels as opposed to energy during the early 1970s, the nation was without any agreement on a conceptual scheme for analyzing energy supply and demand over even the short term. Each of the energy resources had its own nomenclature, data were fragmented and organized around different measures and sets of assumptions, and the information base in the United States was poorly linked to global energy assessments. Stobaugh and Yergin (1979, p. 14) noted that six years after the embargo these still were major difficulties: "We have yet to meet an energy specialist who can keep all the barrels of oil, trillion cubic feet of natural gas, tons of coal, gigawatts of electricity, and quads of energy in his (or her) head. We certainly can't."

Executive and legislative decision makers were thus faced with formulating a comprehensive, systematic national energy policy without a meaningful conceptual framework for understanding energy supply and demand. Most Americans found even elementary comparisons among the various energy resources almost impossible to deal with.

To a large degree, it was the need to bring order and understanding to the energy system that caused the rapid shift to the use of energy models. In 1974 the Federal Energy Administration (FEA) undertook the construction of a computerized model of the United States energy system, called Project Independence Evaluation System, or PIES. This model has been called the most complex and highly sophisticated of the models that were built during the period (Koreisha and Stobaugh, 1979, p. 254). Much of the national energy policy debate was organized around PIES and a plethora of other models in the immediate postembargo period. In fact, however, the models, rather than providing a common information and reference point that could order energy policy debates after 1973, became themselves a source of controversy.

Most of the models produced extremely optimistic assessments of the nation's ability to produce more energy and extremely pessimistic views of its conservation potential. Even though they offered optimistic production estimates, the differences among the various models were sufficient to provide ammunition for contending parties seeking differing policy choices. In the end what in fact happened in the energy system demonstrated that all the models were inaccurate. According to Koreisha and Stobaugh (1979, p. 262):

The use of formal models to provide insight can be beneficial. The dilemma of formal models, however, is that the scientific auras surrounding them encourages those who use the results of models to expect more than this. Such expectations at times are encouraged by model builders who get so enthralled with their own models that they promise much more.

One characteristic that all the models shared was a technological optimism. Thus in developing their models this new group of energy experts had a tendency to downplay social and political factors. Many of the projections derived from the models omitted sociopolitical variables that turned out to be of critical importance in the actual development of American supply and demand (see, for example, Institute for Energy Analysis, 1979). The technological optimism ultimately contributed to disorder and to the heightening of suspicions.

In the end the president and Congress consistently faced a dilemma: when the results of models supported one or another advocacy group's positions, the model projections were vigorously used. When the projections were contrary to such positions, however, advocates attacked the models with equal vigor. The overall effect, therefore, was to contribute to the already widespread uncertainty about the various policy choices.

Impact-Motivated Participants

Concern with the impacts of particular energy resource choices was another major motivation for involvement in energy policymaking. Impact motives largely are based on questions about economic and environmental benefits, costs, and risks. Thus two new sets of participants entered the energy policy arena for impact reasons: short-term economic advocates and environmental and other second-order impact activists. Like the performance-motivated actors, these sets of participants had been active before the embargo. Now, however, they became much more active and influential as the country turned its attention to the new "menu" of energy alternatives and the concomitant second- and higher-order environmental, health, and safety consequences of proposed production and consumption processes and technologies. While certainly not new to the energy scene, both sets of actors increased in numbers and intensity.

On the economic-benefit side of the equation, the key post-embargo difference was the sheer magnitude of interests mobilized by the prospect of an accelerated diversified national energy development program. It was expected that the traditional economic benefits to be made available by energy development—more jobs, higher wages, increased business opportunities—would be large. The growth in federal energy research and development (from just over $1 billion dollars in 1974 to almost four times that amount in 1980) and associated energy subsidies (such as the appropriation of $20 billion for synthetic-fuels commercialization) had tremendous support as pork-barrel politics.

Increasingly sophisticated research on the second- and third-order implications of energy development changed the way

this development was viewed. Fueled by the growing impact-
and technology assessment movement, the number of impact-
concerned actors increased. In addition to the traditional concerns with first-order impacts there were now many actors
concerned about much longer term implications.

Short-Term Economic Advocates

As emphasized earlier, as long as energy supplies were abundant and prices were either stable or declining, the management of energy was left to the stable fuel policy systems. After
1973 the price of petroleum went up tenfold, and the prices
of all other sources of energy increased (though not as much
as oil prices). As will be shown in later chapters, during the
next seven years the efforts by the president and Congress to
formulate a national energy policy repeatedly had to deal with
two issues. One issue was whether it was appropriate for government to allocate energy resources and, if so, how. The other
was whether government should regulate the price of energy
and, if so, how. The pervasive importance of energy to every
other economic interest group in the United States meant that
nearly every potential winner and loser in society became actively involved at one time or another in efforts to influence
the direction of energy choices.

During the periods of shortage interest groups ranging from
resort owners to truckers mobilized to exert influence to gain
a larger share of available fuels. Similarly, owners of independent retail gasoline stations complained that they were
being driven out of business because the large oil companies
were diverting scarce fuel to company-owned stations and
making it unavailable to them. The situation can be characterized as one in which those interests which experienced (or,
in some cases, merely felt they were about to experience) shortfalls reacted by putting pressure on the Congress and the president.

Over the longer term the economic issues triggered by rising
prices were among the most abrasive and difficult imaginable.
Interest groups split depending upon how they perceived higher
prices would affect them. On one side, the traditional energy
industries vigorously advocated deregulation of prices. Energy-
producing states supported this stance, as did many economists and some energy-conservation groups. Also, some groups

advocated decontrol of prices because the ideology they espoused convinced them that this kind of government intervention caused more harm than good. On the other side, consumer states—especially those of the Northeast—representatives of economically disadvantaged groups, a minority of economists who saw deregulation as a trigger for inflation, and many persons who viewed the energy crisis as a creation of the oil industry advocated government price control.

At its roots the pricing controversy was inextricably bound with the larger social controversy about the fairness, equity, and wealth-distribution implications of the energy crisis. Clearly, rising energy prices hurt those below the poverty line the most. Moreover, price fluctuations were major sources of regional strife in the postembargo period. In short, the direct and indirect burdens that accompanied the end of cheap, abundant, and secure energy fell unevenly upon American society, and pressure built for government to distribute these burdens more evenly (see Landsberg and Dukert, 1981). In such an environment it was perhaps inevitable that participation by those with short-term economic vested interests would become highly parochial. Only when it became possible to make a fundamental decision to deregulate energy prices, in conjunction with policy decisions designed to equalize economic impacts of higher prices, such as the windfall profits tax, was it possible to minimize the disorderly participation by transitory special-interest groups.

Environmental and Other Second-Order Impact Interests

The 1970s are frequently characterized as the environmental decade. In truth, that probably is too narrow a characterization of what occurred during the 1970s. The dominant concerns of the 1970s were not just with ecological impacts but with a broader range of second-order or previously unanticipated consequences of many production-technology decisions.

The energy policy debate of the 1970s was strongly affected by two developments. First, a third new group of experts rapidly developed and began participating vigorously in the energy policy struggle. These were scientific-technological experts who focused primarily on gaining an understanding of the second-order consequences of energy developments. These experts sought to integrate environmental and other second-order concerns into the definition of technological performance used in judging new energy processes and techniques. Second, a broad, diverse collection of impact-oriented interest groups became very active. Such organizations as the Sierra Club, the Natural Resources Defense Council, and the Audubon Society became major participants and demonstrated their capacity to sustain the advocacy of their concerns in the presidential-congressional policymaking process.

It is important to distinguish the new group of impact-oriented scientific and technical personnel and understand why they developed. It should be recalled that the traditional scientists and engineers, and even many of the new technical promoters of alternative sources of energy, were interested primarily in developing energy options that would be evaluated on an economic basis. The "new" impact-oriented experts were concerned with economics, but their overriding interest was in understanding the impact of emissions and effluents flowing from existing and proposed energy sources because such understanding was necessary to carry out environmental regulation. Expertise in air and water pollution, waste disposal, and so forth was called for by a whole spate of environmental legislation. With the creation of the Environmental Protection Agency came large-scale federal funding for environmentally oriented scientific and technical work. Later, EPA funds were supplemented by the Department of Energy. With federal funds states, private-sector actors, and a host of nonprofit research organizations developed their own cadres of environmental- and social-impact experts.

The theoretical underpinnings of impact research are nascent, and the introduction of this new body of expertise into the energy policy debate increased substantially the uncertainties that the president and Congress had to confront. The new experts not only raised grave issues concerning the impacts of new options, such as the breeder reactor, but also began raising serious questions about available technologies as well. For example, the environmental consequences of coal-fired electric power plants, previously viewed only as minor liabilities, became much more controversial when they came under the scrutiny of the new environmental scientists and technicians. Thus such problems as acid rain rose to high positions on the national energy policy agenda. And, when these environmental costs were placed alongside the already complex problems of comparing the economic costs of energy options, policy decisions were made more complex by orders of magnitude. By the end of the decade hardly an option existed that had not been identified as having some kind of adverse impact.

Driven by the growing environmental ethic, the proliferating ecological and consumer interest groups, in a manner consistent with American tradition, made liberal use of scientific information that raised questions about the long-term impacts of proposed energy technologies. The posture of some of the new environmental and consumer groups seemed to proceed from the assumption that if there were uncertainties about impacts no action should be taken until the uncertainties were reduced.

The result of the availability of impact information to new interest groups was that the issue of risk became politicized. Nelkin and Pollak (1979) have noted that in earlier times the risks of science and technology were mainly seen as technical problems, not political issues, and were thus the province of experts—experts who shared a consensus. But controversies over the risks associated with nuclear waste disposal or air pollution caused by fossil-fired power plants have called attention to the basic social and political implications of expert-based decisions. Scientific and technical risks are themselves often poorly understood, and the need to define those risks, identify the relevant costs and benefits for society, and develop policy alternatives to mitigate or eliminate risks blurred the line between technical and political decisions. As a result concerns about the risks of technologies frequently translated into questions about political authority and the legitimacy of those making the decisions. In such a situation the pressure for the experts to become involved in the political struggle was irresistible.

The introduction of an energy impact constituency who acquired ever greater sophistication in the use of information provided by the impact experts almost assured that the nation's approach to energy policy would be fundamentally altered. As long as the performance of energy technologies was judged purely in economic terms, the focus could remain on reaching a clear understanding of the physical factors and applying rational standards, however awkwardly, as guidelines for choice. It was possible to believe, under these circumstances, that the way to resolve problems was to rank energy alternatives on the basis of "scientifically validated and fully routinized procedures" (Bereano, 1976, p. 441). Incorporating unanticipated second-order impacts blew this assumption to bits.

After environmental and other kinds of impact concerns became legitimate considerations in the policy process, differing perceptions of the future became an integral part of the energy debate. The technology-assessment movement was only the most visible manifestation of a focus on the higher-order consequences of energy projects. This movement fought for the notion that the distribution of impacts ought to be a major focus of policy analysis. Out of this orientation came a host of new policy institutions, including the congressional Office of Technology Assessment, several university-based interdisciplinary groups, and a large number of analysts committed to the investigation of the "consequences of technology on individuals and groups in society, on the eco-system (the natural environment), on the economy, on other technologies, and to the extent possible, on politics" (Hahn, 1975, p. x).

The president and the Congress were forced to take these concerns into account. For the most part the courts were unprepared to exercise independent judgment about such concerns. Using the requirement in the National Environmental Policy Act for environmental impact statements and various other legislative mandates that drove policy in the same direction, the courts were, however, willing to block actions by federal agencies that had not taken future impacts into consideration. From their earliest days many of the environmental interest groups used the judicial system as a fall-back mechanism when they were unable to achieve their goals through the market or when they felt that the representative (legislative and executive) or bureaucratic institutions were not performing adequately in safeguarding the ecosystem (Garvey, 1975). The courts became, for these impact-oriented interest groups, a mechanism of veto politics. The courts were not major contributors to conflict resolution, but they did give the environmental interest groups powerful leverage in forcing compromises with government agencies and other advocacy groups.

The complications posed for policymaking by the impact-oriented interest groups were worsened by the substantial differences of opinion among the new environmental interests. According to the *Congressional Quarterly* (1982, p. 155):

A major problem facing environmentalism in the 1980s is likely to be harmonization of internal differences as the movement embraces more and more wide ranging concerns. Environmentalists are quite bitterly divided on some of the major policy issues of the day. Conservationist organizations tended to favor Carter's plan for oil price decontrol while groups aligned with urban minority and labor interests preferred rationing and strict public regulation of the energy corporations.

Given the nascent state of impact-oriented research, the fundamental problems of integrating detailed projections of the future into present policy, the tendency to use the courts as a veto-politics body, and the fragmentation and unpredictability of many of the environmental and consumer interest groups, this collection of advocates added substantial complications to the efforts to find compromise solutions to national energy policy throughout the 1970s.

Power-Motivated Participants

The fourth rationale for taking part in energy policymaking was the desire to achieve a certain measure of control over the decision-making process. Much of the expanded involvement in energy policy of recent years has had to do with concerns about how one gains access to the levers of policy and how policymaking institutions might be restructured to incorporate more diverse concerns. Some of the power-motivated actors in energy policy were driven by very traditional domestic and international concerns, but as the following discussions indicate, some of the new players in energy were participating because of a desire to make substantial changes in the sociopolitical system.

Domestic-Policy Actors

If perceptions of economic costs or economic advantages served to mobilize nearly every interest group in the United States, perceptions of power also drove federal executive agencies and

100

congressional committees in the same way. At one time or another following the energy crisis nearly every committee of Congress held hearings on energy issues and sponsored legislation aimed at resolving energy controversies. In a similar fashion nearly all federal agencies and departments proposed actions aimed at overcoming the constraints on energy independence. The increased involvement in energy policy for power purposes was not restricted to the federal government, however. Because performance and impact concerns mobilized a broad cross section of American society on energy issues, regional, state, and local interests also became protagonists in energy debates, many of them for the first time.

Energy was seen for a while as the predominant issue on the national policy agenda. As a consequence, nearly every institution in the federal government structure moved to "gain a piece of the energy action." A large number of federal agencies and committees of Congress became energy advocates with at least one motive in common: to seize the opportunity to gain more important roles in energy policy. The struggle over policy turf frequently represented the same kind of narrow interest characteristic of advocates motivated by economic concerns.

Chapters 8 to 10 investigate the evolution of energy responsibility within federal agencies during the period 1973 to 1980; thus this discussion of the national apparatus can be brief. It should be emphasized here, however, that significant actors were mobilized for power reasons outside the federal government as well. To cite only the most obvious example, the western United States, as the potential source of much of the coal, uranium, oil, oil shale, and geothermal energy in this country, naturally had significant incentives to undertake the development of its own indigenous institutions. To that end the Western Governors' Policy Office was created and became a major actor in the ongoing debate over the merits and deficiencies of efforts to pursue a more aggressive strategy for energy-resource development in the West.

Foreign-Policy Actors

There can be little doubt that many of the fundamental energy issues are international in scope. It was, after all, an international organization, the Organization of Petroleum Exporting Countries, that focused Americans' attention on the need for more comprehensive and better-coordinated global solutions to the complexities of the energy crisis. Following upon the oil embargo of 1973 was a growing recognition that national energy policies did not function in isolation but were part of an interdependent international energy system that was assuming more significance with the growing appreciation that worldwide energy resources were finite (see Willrich, 1975).

With the discovery of oil in the Middle East, that area became a concern in American foreign policy. It was the embargo, however, that made energy an international issue. Energy fuels, particularly oil, had a global market. The embargo signaled an effort by the Arab oil-producing nations to use petroleum as a political instrument. During 1974 and 1975, Secretary of State Henry Kissinger made a vigorous effort to take United States energy decisions in directions that would serve foreign-policy interests. At his initiative the members of the Organization for Economic Cooperation and Development agreed to maintain petroleum reserves, establish mandatory restraints on oil consumption, and set up an international allocation system that would go into effect in the event of another disruption.

Henry R. Nau (1980) has characterized the early perception of the foreign-policy implications of the energy crisis as one of "diplomatic threat." That perception was soon followed by the view that the global energy crisis really was an economic threat, in which higher prices were leading to recessions in the oil-consuming countries. This view was soon replaced in turn by the image of energy issues as institutionally based crises of global proportions—as difficulties between developed and less-developed countries. Finally, energy came to be seen as a global resource shortage, in which the demand for energy was foreseen as outstripping international supply within less than a decade. Throughout these changes in American perceptions of the foreign-policy implications of the energy crisis, attempts were made to fit international energy realities to the needs of American foreign policy. Underlying many of these debates was the issue whether the United States should cooperate with the oil-exporting cartel or try to break it (see De Marchi, 1981).

These changing views produced severe complications for national energy policy during the Nixon and Ford administrations. During the Carter administration foreign-policy concerns once again became primary in the positions taken by the White House. At a summit conference held in 1978 in Tokyo to address economic issues, President Carter made a commitment to reduce United States oil consumption. At the time he made that commitment, it was not obvious that United States demand for petroleum was going to decline. To maintain its credibility, the administration therefore exerted great pressure on Congress to take some kind of meaningful action. For the world as well as for the United States energy had come to the very top of the agenda. Many countries allied with the United States depended on imports at the time of the embargo for almost all of their energy. They saw the United States as the profligate energy user in the world and, therefore, the source of the international energy problem. Energy was thus a central ingredient in foreign policy, and it further complicated efforts to formulate a national energy policy.

Advocates of Sociopolitical Change

As energy became the vehicle for focusing attention on different approaches to public policy, it attracted activists whose motivations were only peripherally related to energy. As noted in the quotation at the beginning of this chapter, energy had become the testing ground for conflict over broader social choices.

If energy was the lifeblood of American society, it was also the area of public policy with the greatest leverage in terms of power. Clearly many advocates saw in energy the lever that could be used to achieve what they perceived to be necessary changes in the fundamental structure and character of American society. Efforts to use energy as the vehicle for pursuing those changes had particularly destabilizing effects on policy. Since the advocates of social-system change were not primarily concerned about energy, they were not consistent participants in the effort to formulate energy policy. Rather, they all too frequently behaved in a guerrilla fashion, becoming involved in hit-and-run fashion.

Perhaps the clearest way to illustrate this point is to discuss briefly four sets of concerns that motivated advocates of social-system change to take part in the struggle surrounding energy policy. They were (1) an effort to reduce materialism in American society, (2) an attempt to expand public participation in

103

policymaking, (3) a drive toward decentralization of decision making and political power, and (4) an effort to control weapons of mass destruction.

The attack on materialism in American society, or at least what many participants came to see as an attack, initially manifested itself in the limits-to-growth argument. Widely read books by such authors as Donella H. Meadows (1972) established a framework for this world view. Out of a growing concern with declining natural resources and growing environmental problems, a body of thought and advocacy developed suggesting that the planet ought to be viewed as a closed system, a "spaceship earth." If man was not to destroy or seriously deform this system, it was argued, it would be necessary to learn to live in a different, less consumptive way. Although in many of the initial formulations this argument was augmented with facts and numbers, it quickly triggered a debate over the character of American society.

The clearest manifestation of the argument appeared in the context of energy conservation policy. In truth, the moralistic tone of the need to conserve energy was reflected in President Jimmy Carter's famous call for changes that he characterized as the "moral equivalent of war." On the opposing side critics of the belt-tightening view of energy consumption argued that it really was a hidden agenda aimed at nothing less than changing the economic and social structure of the United States. Only late in the 1970s did the dialogue about conservation become largely free of this moralistic tone.

The embargo added powerful impetus to another movement, which had its beginnings in the late 1960s. This was the growing dissatisfaction with the way decisions were being made in American society. Doubtless the civil-rights revolution and the Vietnam War influenced this unease. Advocates of increased public participation generally took the view that elites were moving the nation in directions that were not popularly desired and that, in fact, America's democratic system was being eroded.

The drive for public participation proceeded from the assumptions implicit in democratic theory that citizen involvement produces better politics and better decisions. Public participation is believed to be the way democratic societies alter power relationships and legitimize decisions; it serves as a communication device to funnel information into policy systems; it is a means of conflict resolution; and it may be a way to reduce the sense of alienation in the individual citizen (Wengert, 1976). To these general arguments David Orr (1979) has added specific reasons why public involvement should be encouraged in energy decision making: participation may alter the outlook and behavior of the involved citizen and encourage greater awareness of the wider public interest, and involvement by a broad range of parties may promote greater equity in the distribution of the costs and benefits of energy policy by acting as a countervailing influence to offset elite biases and ensure political accountability.

Each of these justifications was used at one time or another by advocates seeking access to energy policymaking. The arguments were particularly helpful to environmental interest groups. As in all other social and political areas, of course, once having gained access to the machinery of decision making, many of the new advocacy groups manifested some of the same elite characteristics and resistance to genuinely open policymaking that they had previously attacked. Unquestionably, however, the pressure for broader participation was very effective. Through legislation and regulations an almost infinite number of opportunities developed for citizen involvement through the requirement of environmental impact statements, intervener mechanisms in siting processes, and so on. With all these openings energy decision making was altered by intermittent participation by advocates of varying causes. Leaving aside the question of whether public participation was beneficial or detrimental, the opportunity for broad citizen involvement appears to have been a significant factor in slowing the pace of energy-related decisions at almost every level.

Motivated by the same concerns about the functioning of American democracy, another group of energy advocates sought to counter what they saw as the increasingly undemocratic character of this technological society by pursuing much broader social and economic changes. Although the argument is framed in vague terms, the widely discussed call by Amory B. Lovins (1977) for the pursuit of a "soft energy system" was driven by the perceived need to avoid public alienation, which was seen as resulting from the ever-growing technology and its requirement for increasingly concentrated decision-making power.

105

Lovins and Howard T. Odum (1971) are representative of a group of advocates who sought access to energy policy for power reasons. If technology has a great deal to do with the shape and direction of politics in a democracy, then a highly technical arena such as energy inevitably acquires major significance.

Once these motivations became important, energy policy no longer was exclusively the turf of people with a direct stake in energy performance or impact. Now the energy policy system included a number of advocates who were involved for what had been earlier thought of as "vicarious" motives. That is, the fact that energy decisions were so intimately linked to society as a whole increasingly was used as a rationale by new actors intent upon using energy policy to attain nonenergy goals and objectives.

The clearest example of vicarious participation in energy matters occurs in the nuclear power policy field. For many parties-at-interest in this area, the performance or impact questions raised by the nuclear fuel cycle were of little significance beyond the goal of gaining political leverage. In the words of Lovins (1977, p. 56):

. . . if nuclear power were clean, safe, economic, assured of ample fuel, and socially benign per se, it would still be unattractive because of the political implications of the kind of energy economy it would lock us into. But fission technology also has unique sociopolitical side effects arising from the impact of human fallibility and malice on the persistently toxic and explosive materials in the fuel cycle. For example, discouraging nuclear violence and coercion requires some abrogation of civil liberties; guarding long-lived wastes against geological or social contingencies implies some form of hierarchical social rigidity or homogeneity to insulate the technological priesthood from social turbulence; and making political decisions about nuclear hazards that are compulsory, remote from social experience, disputed, unknown, or unknowable may tempt governments to bypass democratic decision in favor of elitist technocracy.

This is not to say that each of these concerns is not a legitimate issue for public policy in this country—they clearly cut to the essence of what democracy is all about. The point is that the search for resolutions to these issues in the context of formulating an energy policy must inevitably encumber and constrain the policymaking process. The broader social issues so extended the boundaries of the energy policy debate that every aspect of society had to be taken into account. And it grew worse. Lovins (and many others who had similar viewpoints) focused on energy policy levers, but for many of the vicarious participants in the energy debate in the early and middle 1970s, energy simply was a vehicle for advocating a different view of what the world should be.

Clearly, controversies such as nuclear reactor siting became, in some way not fully understood to this day, linked to protests against the Vietnam War, Watergate, and other social and political issues that had little or nothing to do with the conduct of energy activities. Moreover, this vicarious participation tended to take place at the most sensitive points in the energy-resource-development process. In the nuclear power field electric utilities were subjected to public protests against the location of reactors and other "noxious" facilities. Similarly, when gasoline shortages developed, the retail-gasoline sector of the petroleum industry received the brunt of the protests. This was no coincidence. If playing out one's world view on the energy stage is to have an impact on the polity as a whole, it is important to hit the system where it is most vulnerable.

In retrospect one thing is clear. All these disruptions deeply affected the national energy policy debate. The rapid formulation of a national energy policy was made impossible.

In close association with the participants described above were other interests involved in energy policy struggles in the 1970s that had fairly specific and discrete policy goals. These were the opponents of the arms race and weapons proliferation. Again, the nuclear energy field provides the clearest example. Some of the participants involved in the antinuclear movement were motivated by concerns about the nuclear-weapons race. For them nuclear energy provided the most convenient, effective arena for making their concerns manifest. For example, opposition to the reprocessing of nuclear fuel was high on the agenda of those motivated by concern about the proliferation of nuclear weapons. In sympathy with these concerns, the Carter administration decided to place a moratorium on reprocessing activities, and this decision was explained primarily in terms of nonproliferation.

Thus did the number of participants in the energy policy

system expand dramatically during the decade of the 1970s. As later chapters illustrate, it was only at the end of this period that an elementary consensus about the boundaries of that system and its community could be constructed. The following chapters focus on another set of complications that stood in the way of the rapid development of an energy policy: uncertainties over energy resource options.

CHAPTER 5

Uncertain Alternatives: Oil, Natural Gas, Coal

Our nation's energy problem is not a technological one. We do not lack technologies and resources that can provide us with desired quantities of energy in familiar forms. We have, or almost certainly can develop, technologies to fit a wide variety of specifications for energy supply, resource use, environmental protection, and human safety. — Thomas J. Wilbanks, *Building a Consensus About Energy Technologies*, ORNL, September, 1981

The earliest and most persistent proposals for overcoming the energy crisis called for finding and developing new sources of energy. Following the oil embargo advocates began proposing new sources of energy (oil shale, tar sands, organic wastes, geothermal power, solar power, and conservation), as well as many new technologies for using the conventional sources (for example, synthetic fuels ("synfuels") from coal, fuel cells, and magnetohydrodynamics).

The dilemma faced by the president and the Congress was twofold. As discussed in chapter 4, many actors and interests were advocating conflicting special interests. In addition, many complex physical activities were being advocated as viable options. A continuing problem for the president and Congress through the 1970s was the lack of consensus among experts on which of the physical options should receive priority.

With minor exceptions none of the newly proposed sources of energy was available. A central challenge for the president and Congress after 1973 was to decide what activities should be included in the energy sector and in which subsector they should be placed, (1) available, (2) potential, or (3) theoretically possible.

As described earlier, available activities are those identified as off the shelf. The major distinguishing characteristics of available activities are that reliable performance estimates can be made about what they will cost and how long it will take before they can produce. Potential activities are characterized by substantial disagreement among scientific-technical experts about what they will cost and how long it will take before they can produce. This technical uncertainty leaves considerable leeway for conflicting interpretations about costs, risks, and benefits and therefore increases the difficulty in determining who will pay for or enjoy them. Theoretically possible activities are those on which scientific-technical experts agree that no reliable calculation can be made about costs or when they can produce.

Since few of the major new sources of energy proposed as answers to the energy crisis were in the available subsector, the president and Congress were faced with choosing among sources of energy about which there was performance uncertainty.

Chapters 5 to 8 describe the range of energy options and characterize them in terms of their level of technical development: available, potential, or theoretically possible. An appreciation of the range, complexity, and uncertainty of the choices on the energy "menu" is central to understanding the difficulties associated with formulating an energy policy between 1973 and 1980.

The twelve possible sources of energy that were the focus of the conflict resolution process undertaken between 1973 and 1980 are shown in table 5.1. The various major activities included within each energy resource option are shown by subsector. These activities are described in the chapters that follow.

The task undertaken in these chapters is inherently difficult, and the reader should be alerted to two limitations. First, the classification of the major energy alternatives in terms of their stage of development (available, potential, and theoretically possible) represents a determination made by the authors based on a review of the literature and interviews. Clearly disagreements exist among experts about the stage of development of certain energy activities, and some readers will disagree with our classification of specific activities (most objections will come from those who argue that activities classified here as potential are in fact available). Consistent with the classifica-

Table 5.1. Resource options by subsector of the energy sector

Option	Available	Potential	Theoretically Possible
Crude oil	Conventional activities	Deepwater and Arctic production Subsea production systems Enhanced recovery	
Natural gas	Conventional activities LNG systems	Tight gas production	
Coal	Mining Benefication Transport Burning	Liquefaction Gasification Improved solids	
Nuclear	LWR	HTGR LMFBR Reprocessing Waste storage	Fusion
Hydroelectric	All conventional hydro options	Tidal power	
Oil shale	Mining Transport Upgrading	Surface retorting In situ retorting	
Tar sands	Mining Surface processing	In situ processing	
Geothermal	Vapor-dominated	Liquid-dominated	Geopressured hot dry rock
Organic waste	Burning	Liquids Gases	
Solar	Low-temperature collectors Wind power Biomass production Alcohol	High-temperature concentrators Ultrahigh-temperature concentrators Photovoltaic cells Ocean thermal gradients Nonalcohol liquids and gases	
Electric power	Boilers Stack gas cleaners Gas turbines Transmission	Fluidized beds Binary cycles Fuel cells Storage Cryogenic transmission	
Conservation	All consumption sectors		

tion system, when we found any disagreement among experts, we placed the activity in the potential subsector.

Second, energy activities are classified solely on the basis of the stage of development achieved in the United States. In the development of certain energy activities the United States clearly lags behind other countries. For example, France is ahead of the United States in the development of the nuclear breeder reactor. We have restricted ourselves to developments in the United States for two reasons. One is that technologies that are economically and environmentally acceptable in other countries are not necessarily acceptable in the United States. The other is that we had only limited access to information on development in other countries.

Although the stated limitations are real, they do not seriously affect the purposes of the following chapters, which are twofold: (1) to provide a broad grid map of the range and complexity of the energy options being advocated since the embargo and (2) to provide a sense of the uncertainty associated with proposed new energy options. In sum, it is clear that both the variety of energy options and the high level of performance uncertainty associated with many of them made the task of the president and Congress in choosing the nation's future energy sources very difficult.

OIL

Oil was first discovered in the United States in 1859, and in 1951 it replaced coal as the nation's primary energy source. Oil's rapid takeover of the United States energy market occurred because it was relatively inexpensive, easily transportable, and the most flexible of the energy sources. Because of these characteristics it is commonly considered a substitute for all other fuels.

At present oil is the nation's transportation fuel, and its takeover paralleled the development of the present transportation system. Oil has also become an important raw material in the development of a broad-based chemical and product industry.

Until 1948 the United States was a net exporter of oil. After 1948, United States consumption increasingly exceeded domes-

tic production, even though United States oil production increased until 1970 (U.S. Congressional Research Service, 1980,
pp. 126, 130). At the time of the oil embargo the United States
was consuming approximately 17 million barrels of oil a day
and was importing approximately 6 million barrels a day (U.S.
Congressional Research Service, 1980, p. 126).

Oil Resource-Reserve Estimates

The nation's failure to anticipate and prepare for a domestic
oil shortage resulted in part from a pattern of optimistic estimates of oil resources that continued into the 1970s. For example, the National Petroleum Council estimated in 1971 that
there were still 810 billion barrels of oil resources in the United
States (National Petroleum Council, 1971, p. 38). Under the
standard rule of thumb one-third of the 810 billion barrels
would be recoverable. To underline the rapidity of change in
resource estimates following the embargo, in 1974 a major oil
company estimated that only 88 billion barrels of undiscovered,
recoverable oil were available to the nation (Gillette, 1974, p.
128). Two general labels are used in discussions of yet-to-be-
produced oil. One label is resources; the other, reserves. Resources refer to all of the oil estimated to be in the ground.
Reserves refer to known resources that are technically and
economically available at a given time.

No technology is available to determine positively whether
oil is in a particular location in the ground before drilling.
From year to year the quantity of reserves is being reduced by
production and added to by drilling. Estimates vary accord-
ingly. In 1974 the nation's reserve estimate was 39 billion
barrels of oil (U.S. Congressional Research Service, 1980, p.
364). In 1978 the reserve estimate had dropped to 33 billion
barrels, and by 1981 to 29 billion barrels (U.S. Department
of Energy, 1982, p. 44). Since the embargo the United States
has extracted more oil than it has found every year except
1980 and 1981, when new reserves about equaled production
(Bureau of National Affairs, 1982b, p. 930). Those years were
exceptions because the increased price of oil brought about
markedly increased investments in drilling, but they represent
no real turnaround in the nation's projected-reserves picture.

Location of Oil Resources

Most domestic oil comes from the western half of the contiguous forty-eight states, the second-largest portion comes from the North Slope of Alaska, and the third major portion comes from the Gulf of Mexico. Three areas are thought to offer the major potential for large new discoveries of oil: the outer continental shelf, onshore Alaska, and the Western Overthrust Belt, which runs along the Rocky Mountains.

The federal government is the owner and the Department of the Interior the primary manager of most of the highly prospective oil lands. Federal land policy, therefore, has been a major focus of the energy policy debate that has gone on since 1973. In general, the controversy over federal lands has pitted those concerned primarily with environmental impacts who urge exploration at a more deliberate speed because of fear of environmental damage against those concerned with greater domestic energy self-sufficiency who wish to see exploration occur on a very expedited basis.

Activities

Oil activities are normally divided into four categories: exploration, development and production, transportation, and refining.

Exploration

Exploration activities include both surveys and exploratory drilling. Survey methods include passive measurement techniques that aid in identifying irregularities in subsurface geology, changes in the earth's magnetic field, and local variations in the earth's gravity that reveal potential geological traps in which oil may have accumulated, as well as finding natural oil seeps. Active surveys use both core drilling and, most frequently, seismic surveys. The technology of seismic surveying and interpretation has steadily advanced and has given the industry increasing confidence in its results.

Exploratory drilling is undertaken to reach particular geological formations believed to contain oil or gas. Offshore exploratory drilling employs the same techniques as onshore drilling except that a platform must be provided to support the offshore drilling rig. There are four basic types of offshore exploratory drilling platforms, all mobile: barges, jack-ups, drill ships, and semisubmersibles. Barges are normally used in shallow, protected waters. Jack-ups (figure 5.1) are platforms with retractable legs that can be lowered to the ocean floor to lift the platform out of the water. Drill ships (figure 5.2) are used for exploratory drilling in deep water. Semisubmersible rigs (figure 5.3) are floating platforms with most of the flotation submerged. Semisubmersibles have excellent stability in severe weather.

In recent years the oil industry has demonstrated its capacity to test all kinds of geologic structures that are presently thought to have oil-producing potential. In particular, the industry has developed the capacity to drill below 30,000 feet. Similarly, it has demonstrated that it can carry out exploratory drilling in all the water depths likely to be leased in the foreseeable future. All oil-exploration options fall in the available subsector.

Development and Production

Development and production activities can be grouped into four categories: (1) drilling, (2) completion, (3) processing, and (4) enhanced recovery.

Drilling. Development drilling is carried out in the same manner as exploratory drilling. Offshore development wells are usually drilled from fixed platforms. Each platform normally contains a number of wells that are directionally drilled to different parts of the reservoir. Most fixed platforms are constructed of steel, though in recent years some concrete platforms have been constructed. Currently fixed platforms operate in slightly over 1,000 feet of water.

Completion. Wells are completed by setting casing pipe in the hole with cement and installing tubing (pipe) to carry the produced oil to the surface. Where necessary, safety valves are installed to ensure against blowouts. Crude oil is delivered to the top of the well either by natural pressure from the reservoir or by some form of artificial lift, such as a pump or the injection of gas into the reservoir. In a limited number of cases

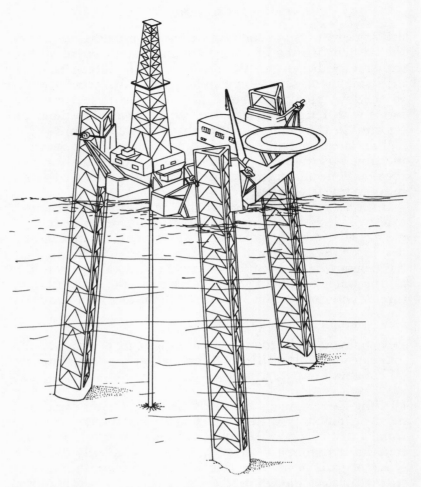

Fig. 5.1. Jack-up offshore drilling rig. Sketch from Esso Production Research Company.

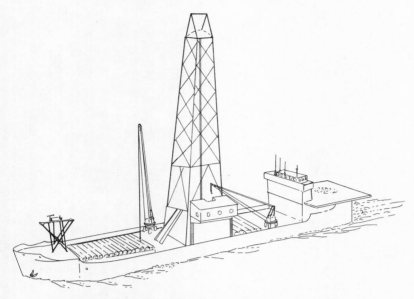

Fig. 5.2. Drill ship. Sketch from Offshore Company.

offshore wells are completed on the ocean floor with what are called subsea well heads. Production activities fall in the available subsector.

Processing. Once it is out of the well, oil is processed in the field to remove natural gas, saltwater, sand, and other impurities, and then quantity is measured before transport. Processing is an available activity.

Enhanced Recovery. When the natural flow of oil from a reservoir diminishes, additional oil may be recovered through the use of enhanced recovery techniques. The most widely used technique is called water flooding (figure 5.4). This technique consists of pumping water down selected wells to wash or push oil from the reservoir into producing wells. Other enhanced-recovery techniques employ chemicals, heat, or gases to remove oil that would not naturally flow from the reservoir. The rule of thumb is that only about one-third of the oil in the ground

117

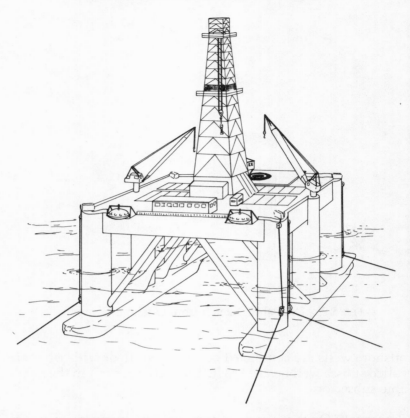

Fig. 5.3. Semisubmersible offshore drilling rig. Sketch from Off-shore Company.

is normally produced without enhanced recovery. Water flood-ing is an available activity. Most other enhanced-recovery tech-niques fall in the potential subsector.

Potential Production Activities. Because rigid, bottom-founded platforms become uneconomic at some water depths, the oil industry is designing and testing a wide range of offshore pro-duction platforms. The new designs fall into two general cate-gories. Platforms in the first category float, attached to the bottom with either cables or a flexible coupling. The objective is to relieve the tremendous pressure generated by a very deep

118

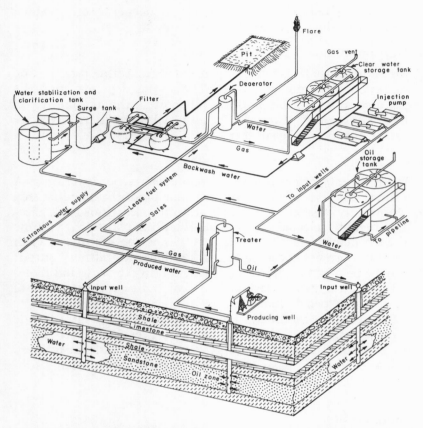

Fig. 5.4. Waterflood enhanced-recovery system. From U.S. Bureau of Mines, 1970, p. 26.

column of water. In designs of the second category, both production and processing facilities are on the ocean bottom. These are called subsea production systems. The new concepts fall in the potential subsector, but they appear very close to being available.

Similar platform experimentation addresses the unique problems of drilling in the Arctic, where the tremendous pressure generated by ice requires special designs. At present gravel islands are used in shallow Arctic waters: such activities fall in the available subsector. Experiments are being made with

119

several alternative designs for use in deeper Arctic waters; they fall in the potential subsector.

Transportation

Oil is presently transported by tank trucks, railroad tank cars, tankers, barges, and pipelines. The choice depends upon the distance traveled, the product transported, the availability of alternatives, and, of course, the cost. All these alternatives are within the available subsector.

Refining

The final category of petroleum-related activities is refining. A refinery converts crude oil into various products. The first step is normally separation by distillation into fractions (e.g., gasoline, fuel oil) selected on the basis of boiling points, the volume of each determined by crude type. Since the volume of each fraction may not conform to market demand, refinery output may be changed by splitting or rearranging the original molecules. Refineries are designed to deal with expected kinds of crude and expected markets.

Refineries are modified in response to changes in either crude supply or market. Like the other activities in the oil sector, incremental and regular improvements are made in refinery processes, but the basic technology falls in the available subsector.

Summary

The vast majority of activities in the oil sector are in the available subsector. Some techniques of enhanced recovery, and deep-water and Arctic production platforms and subsea production systems fall in the potential subsector. There is a consensus that the technology either is available or can be made available with improvements to produce any large quantities of oil that may be discovered in the future. The energy crisis of 1973 did not identify the need for new activities for the production of crude oil, and the oil policy sector remains much today what it was before the embargo. The major uncertainty is how much oil is still available for discovery in the United States.

NATURAL GAS

The rapid increase in the consumption of natural gas in the United States has paralleled that of oil. Before World War II gas was an important source of energy primarily in those regions of the country where it was produced. In 1947 the conversion of two liquid pipelines from the Southwest to the East Coast to gas pipelines heralded the beginning of a major construction program that by 1970 extended the gas-transmission network to all the contiguous forty-eight states.

At the end of 1981 the United States had consumed something over 600 trillion cubic feet (TCF) of natural gas (U.S. Department of Commerce, 1979, p. 602; U.S. Department of Energy, 1982, p. 105). Consumption since 1981 has been slightly less than 20 TCF per year. The rapid growth in the use of gas is revealed by the fact that 80 percent of the cumulative consumption through 1981 had occurred in the post–World War II period. One-third of the total was consumed between 1971 and 1981.

Gas consumption grew rapidly for four reasons. First, gas is reliable. Second, it is particularly clean and convenient. Third, as a result of regulation by the Federal Power Commission, it has been consistently cheaper than oil. Fourth, the nationwide pipeline network made it available from coast to coast. Natural gas was providing approximately 30 percent of the nation's energy in 1973 and has remained at that level in the years since (U.S. Congressional Research Service, p. 20; U.S. Department of Commerce, 1981, p. 577).

Gas Resource-Reserve Estimates

The general resource-reserve picture for natural gas parallels the picture for oil. Throughout the 1960s natural-gas policy was based on optimistic estimates of the quantity yet to be discovered and produced. In the early 1970s those optimistic estimates began to be whittled down (Wildavsky and Tennenbaum, pp. 328–29).

Like oil, gas in the ground is divided into two categories: resources and reserves. Again, resources refer to all of the gas estimated to be in the ground; reserves, to be discovered gas technically and economically producible. There were sufficient reserves in 1981 to support the nation's existing level of production for approximately ten years (U.S. Department of Energy, 1981, p. 10).

The nation's gas-supply situation is more optimistic than that for oil for two reasons. First, many known gas reservoirs had not been tapped before the late 1970s because production was uneconomic when gas prices were held at a low level by federal regulation. Second, pressures and temperatures increase at lower depths in the earth, and these conditions favor gas formation over liquids. Since deep-gas drilling is a relatively recent development, deep-gas resources have yet to be produced in major quantities. No similar prospects appear likely for new oil production. When deregulation began in 1978, vigorous efforts to find and develop gas reservoirs plus reduced demand created an almost overnight surplus.

Formal estimates of conventional gas reserves in the United States are in the range of 200 trillion cubic feet (TCF; U.S. Department of Interior, 860, p. 2). This figure includes 34 TCF in Alaska (U.S. Department of Interior, 860, p. 25). In addition, there may be another 200 TCF in inferred reserves (gas believed to be in producing fields which has yet to be proven). Finally, the most optimistic estimate is that there may be as much as 655 TCF of undiscovered conventional gas resources. Depending upon whether one is optimistic or pessimistic, the United States has enough conventional gas to last for between 20 and 57 years, at present consumption rates.

Location of Gas Resources

Potentially large new conventional gas resources are in the same areas identified for oil, specifically the outer continental shelf, onshore Alaska, and the Western Overthrust Belt. The deep Anadarko Basin, in western Oklahoma, may contain very large additional resources.

Large unconventional gas resources have been found in the Rocky Mountain states Wyoming, Utah, and Colorado. Estimates are that 600 TCF of natural gas are located in tight (low-permeability) sand and shale formations in those states. These resources as well as potentially large gas resources in Devonian shales that lie along the Appalachian Mountains are not normally identified in gas-resource estimates because as yet no technically feasible methods have been demonstrated for fracturing tight formations so that the gas can flow at economically justifiable rates.

Activities

Natural-gas activities normally divide into three major segments: exploration, development and production, and transportation. Some gas is transported in liquefied form.

Exploration

Exploration for gas is similar to that described for oil. An advanced seismic technique called "bright spot" in some cases raises the confidence in seismic exploration for gas above that which exists for oil. Gas-exploration technologies fall in the available subsector.

Development and Production

The equipment and techniques used to complete natural-gas wells are also similar to those for oil. Processing technologies, however, differ significantly from those for crude oil. The first phase in processing is the separation of the gas from oil, water, sand, and other undesirable compounds. Next natural-gas liquids (e.g., propane, ethane) are separated in natural-gas processing plants. Finally, before transport through pipelines, it is sometimes necessary to compress the gas to ensure efficient movement. Cleaning, processing, and compressing activities may be simple or complex, depending on the character of the gas reservoir and the compounds and other undesirable components produced in the gas stream.

The activities associated with the development and production of conventional natural gas fall within the available subsector.

Tight Formations. For many years both the industry and the federal government have been experimenting with technologies that would free gas locked in tight formations. Massive

123

hydraulic fracturing and other fracturing techniques are being investigated.

Some years ago underground nuclear explosions were tried as a means of stimulating production, but they are not presently considered viable for technical, environmental, and political reasons. One group has estimated that if well-stimulation technologies could be effectively employed as much as 40 to 50 percent of the tight gas could be recovered (Kash et al., 1976, p. 153). The technologies for stimulating the production of gas from tight formations fall in the potential subsector.

Transportation

Most domestic natural gas is transported through pipelines. The technology for laying and operating gas pipelines both onshore and offshore is well established and falls in the available subsector.

Liquefied Natural Gas

The United States is presently both importing and exporting limited amounts of liquefied natural gas (LNG). LNG systems have the following components: a liquefaction plant, including storage and loading facilities; LNG tankers; and unloading, storage, and regasification facilities at the import point (see figure 5.5).

Liquefaction plants must have the capability of refrigerating gas to a temperature of −259 F. Since the liquefaction process is continuous, heavily insulated storage facilities are necessary to contain the gas at very low temperatures.

Liquefied gas is transported in specially designed LNG tankers. The tankers have insulated tanks similar to those used for onshore storage. LNG tankers can carry approximately 70 percent of the energy capacity of oil tankers of the same size.

At the import point the liquefied gas is pumped into storage tanks and from the storage tanks is run through a regasification facility in which the LNG is vaporized by passing it through heat exchangers. Regasification occurs at pressures up to 1,200 pounds per square inch (psi), which is sufficiently high for direct introduction into conventional natural-gas pipelines.

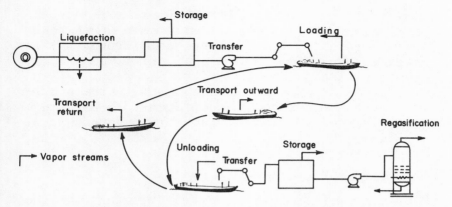

Fig. 5.5. Integrated liquid-natural-gas operation. From Bodle and Eakin, 1971, p. 5.

LNG has been transported by tanker for nearly twenty years and falls within the available subsector.

Summary

With the exception of the technologies necessary to free gas in tight formations, all the activities associated with supplying natural gas are in the available subsector. The major uncertainty is how much gas remains to be discovered.

COAL

From the 1880s until shortly after World War II, coal was the primary source of energy in the United States. Coal, however, is the least convenient fossil fuel, and after World War II oil and natural gas replaced it in a major portion of the energy market. The decline of coal as an energy source can best be illustrated by noting that in 1940 coal supplied 52 percent of the nation's energy but by 1973 supplied only 18 percent (U.S. Department of Commerce, 1979, p. 600).

During the period of decline diesel-powered train engines were substituted for coal-fired locomotives, and in residences

125

gas furnaces were substituted for coal furnaces and stoves. The major growth area for coal has been in steam electric power plants. Between 1960 and 1976 annual utility demand for coal increased from 176 million tons to 596 million tons (U.S. Department of Energy, 1982, p. 127).

Although it appears that the utility market for coal will increase somewhat, any sudden surge in demand will likely come from its use in synthetic-fuels plants or from a growing export market.

Coal Resource-Reserve Estimates

Unlike oil and natural gas, there is little uncertainty about the supply of coal in the United States. The total coal resources are estimated to be as high as 4 trillion tons (U.S. Congressional Research Service, 1980, p. 579). Although at present only 438 billion tons are classified as demonstrated reserves, those reserves would supply the total energy needs of the United States at the 1981 level for nearly one hundred years (U.S. Congressional Research Service, p. 580).

Coal occurs in most parts of the United States, 90 percent of it in four "provinces": Eastern, Interior, Northern Great Plains, and Rocky Mountains (see figure 5.6). Approximately 54 percent of the nation's demonstrated coal reserves lies west of the Mississippi, primarily in the Northern Great Plains and Rocky Mountain provinces. It is estimated that approximately 60 percent of western coal is owned by the federal government, with an additional 20 percent dependent on complementary federal coal for economical mining because of fragmented and intermingled private-federal ownership (U.S. Congress, Office of Technology Assessment, 1981, p. 3).

The two major western coal provinces have the following characteristics: (1) much of the coal is near the surface and susceptible to strip mining, (2) competition for surface-area usage is relatively low, (3) the federal government controls most of the coal lands, (4) the coal is low in energy value per unit of weight and low in sulfur content, and (5) water resources are scarce.

Most coal east of the Mississippi is privately owned. In general, eastern coal has a higher heat value and sulfur content

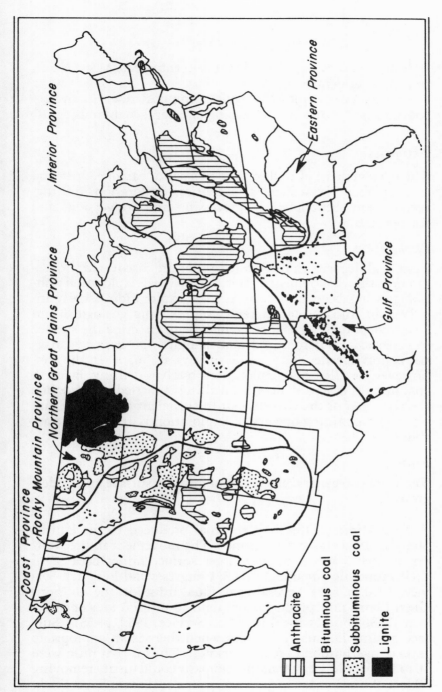

Fig. 5.6. Distribution of United States coal resources. From U.S. Bureau of Land Management, 1974, 1:47.

Coast Province

Rocky Mountain Province

Northern Great Plains Province

Interior Province

Eastern Province

Gulf Province

Anthracite

Bituminous coal

Subbituminous coal

Lignite

than western coal, and much of it is in thin beds. Most eastern mining is underground mining.

The key policy questions concerning coal reserves and resources revolve around the leasing of federal coal lands.

Activities

The development of coal for use as a solid, gaseous, or liquid fuel involves six sequential activities: exploration, mining, reclamation, beneficiation, processing and conversion, and transportation.

Exploration

Most coal deposits in the United States are discovered. Consequently, the development of sophisticated coal-exploration technologies has been unnecessary. Coal exploration usually involves an initial stage of reviewing existing geological and geophysical data. A next stage may be to examine the surface to detect coal outcrops and take samples. On occasion magnetic, gravimetric, and seismic surveys are made. In the end, however, detailed coal exploration involves drilling. Before a coal mine is designed, detailed drilling is carried out to define the character of the coal deposit in sufficient detail to ensure an effective mining operation. Exploration is an available activity.

Mining

There are two general categories of mining: surface and underground.

Surface Mining. Over half the coal produced in the United States in 1980 was surface-mined. The movement in the western United States is to ever-larger surface mines because of their economic advantages (U.S. Congressional Research Service, 1980, p. 538). Several individual mines in the northern Great Plains now produce more than 10 million tons of coal a year (U.S. Congressional Research Service, 1980, p. 551). Surface mining is usually pursued when there is an appropriate ratio of overburden to seam thickness. Thus if that ratio were 10 to 1, it would mean that the company could justify removing 50 feet of overburden to gain access to a coal seam 5 feet thick.

The ratio of overburden to coal seam has been growing in recent years, and some mines producing highly valued metallurgical coal have 40 to 1 ratios.

There are two major types of surface mines, contour and area mines. Contour mining is normally practiced in hilly or mountainous terrain. By this method the overburden is removed from the slope to create a flat excavation, or bench, which is flanked by a vertical highwall on one side and a downslope spoils pile on the other. The exposed layer of coal is then mined (see figure 5.7).

Area mining is pursued on flat terrain. Area mines are opened by excavating a trench to expose the coal deposit. As succeeding cuts are made, the overburden is piled into the cut from which the coal has been mined (see figure 5.8).

Underground Mining. Two methods are used in underground mining, the room-and-pillar and longwall methods. The coal deposit is reached by digging or boring a vertical shaft or a horizontal or slanting tunnel. In room-and-pillar mining, a passageway is excavated through the coal seam. From this passageway rooms are mined out, portions of the coal being left in place as pillars to support the strata overlying the room (see figure 5.9). Room size depends on the character of the strata being mined.

In conventional room-and-pillar mining, a cutting machine makes a slice into the seam. Blast holes are drilled, and the coal is fragmented by blasting. The coal fragments are then loaded and hauled away. In continuous mining, a single machine performs the cutting, loading, and initial transporting operations. As much as 50 percent of the coal in a room-and-pillar mine may be left in place as pillars.

In longwall mining, a sheering drum moves back and forth across the working face of the coal seam between two access passages. The sheered coal drops onto a conveyor, which moves it to the transportation system. The roof of the area immediately behind the mining machine is supported by hydraulic jacks that are moved forward as the mining operation advances. As the jacks are moved, the roof in the area from which the coal has been mined is allowed to collapse (see figure 5.10). The major advantages of longwall mining are the higher percentage of coal removed and the lower labor costs. On the other hand,

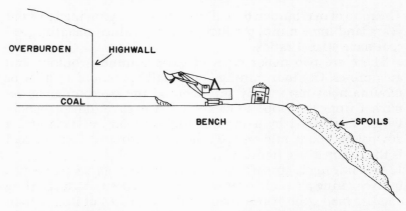

Fig. 5.7. Contour mine. From National Petroleum Council, 1972.

longwall mining is more capital-intensive than room-and-pillar mining. All these mining activities fall in the available subsector.

Reclamation

Both surface and underground mining have significant physical impacts. They disturb the surface, produce wastes that require disposal, and frequently affect water resources by exposing materials that produce acids when dissolved in water. All these impacts pose reclamation problems.

Surface-Mine Reclamation. For both contour and area mines the major reclamation problem is surface disruption. Comprehensive reclamation programs include restoring the premining topography, replacing the topsoil, and revegetating the land. Comprehensive reclamation is much more difficult for contour mines than it is for area mines. Reclamation, however, is now required for all surface mines by federal legislation (the Surface Mining Control and Reclamation Act of 1977).

Underground Mine Reclamation. Reclamation of underground mines can be much more complex than that of surface mines. Reclamation aims at mitigating three impacts:

1. Waste piles that develop around mines. Sometimes these

130

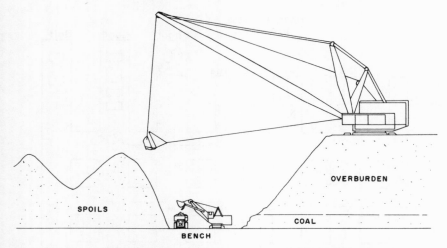

Fig. 5.8. Area mine. From National Petroleum Council, 1972.

waste piles serve as dams. Water impounded by such a pile and released when the pile gave way produced a disastrous flood at Buffalo Creek, West Virginia.

2. Acid water that results when water passes through acidic materials produced in connection with mining. Acids can pollute surface water and underground water in the area of the mine.

3. Subsidence that occurs when the surface overlying underground mines collapses. The pockmarked area that is left is severely limited for any subsequent use.

Reclamation aimed at mitigating these impacts can be very complicated. Like surface-mine reclamation, however, such reclamation must be carried out by federal law.

Although improvements are being made in both surface and underground mining and the associated reclamation, no new activities appear to be on the horizon. Therefore, reclamation activities fall in the available subsector.

In-Mine and Near-Mine Transportation

Within surface mines coal is usually transported by either trucks or conveyors. In underground mines coal is transported

131

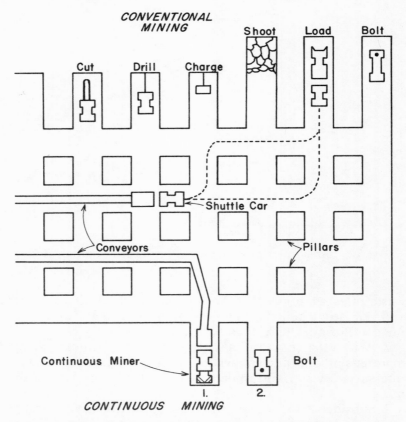

Fig. 5.9. Alternative methods of room-and-pillar mining. From Science and Public Policy Program, University of Oklahoma, 1975, p. 1-29.

by conveyors or rail shuttle cars. In a few mines, pneumatic or slurry pipelines are being used.

Development of in-mine transportation has not kept pace with the development of excavation machinery, but no identifiable technologies of a distinctly different kind are available. In-mine transportation thus involves a series of activities that fall within the available subsector.

Beneficiation

Coal is frequently beneficiated before it is used. Beneficiation

132

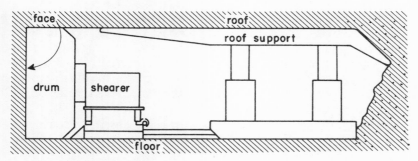

Fig. 5.10. Section view of longwall mining. From Gouse and Rubin, 1972, 3:24.

may involve any or all of the following steps: (1) crushing and screening to desired maximum size, (2) cleaning to remove dust, noncoal materials, and pollutants (primarily sulfur), and (3) drying to prepare coal for shipping or use. Beneficiation technologies fall in the available category.

Processing and Conversion

Coal can be used raw, processed to improve its qualities as a solid fuel, or converted into either gaseous or liquid forms. The technologies for burning coal, as well as for producing gaseous, liquid, or improved solid fuels from coal, are at varying stages of development, as described below.

Coal Combustion. In large facilities most direct coal combustion consists of blowing ground coal into a combustion chamber. This technique gives improved efficiencies over burning the coal on a grate and is now an available technology. Several other combustion processes are being investigated with the dual aim of improving combustion efficiency and mitigating pollution. The best known of these techniques is the fluidized bed, where finely sized coal, mixed with other material, is kept in a boiling state while it is burned by gas passed through it. Fluidized beds and other experimental combustion techniques are in the potential subsector.

Gasification. Gaseous fuels of low-, medium-, or high-energy content can be produced from coal. Low-Btu gas has a heat content of up to 200 Btu's per cubic foot. Medium-Btu gas has a heat content ranging from 300 to 650 Btu's per cubic foot. High-Btu gas is in the range of conventional natural gas, that is, above 900 Btus per cubic foot. Low- and medium-Btu synthetic gases are produced in a two-stage process involving preparation and gasification. A third stage, called methanation, is required to produce high-Btu gas.

Three elements are needed to synthesize gas from coal: carbon, hydrogen, and oxygen. Coal is the source of the carbon. Steam is the source of the hydrogen, and either air or pure oxygen is the source of the oxygen. Heat can be supplied either by directly combusting the coal and oxygen inside a gasifier or by introducing hot pebbles or ceramic balls heated in some external unit.

Various gasification processes are being tested. Each method involves trade-offs. The goal is to produce synthetic gas at the lowest price. One set of choices involves whether to use hydrogen or steam as the source of the hydrogen, air or oxygen as the source of the oxygen, and direct or indirect heat. Another set of choices of an engineering kind involves whether the gasification system is operated at a high pressure or at atmospheric pressure. Finally, three types of bed systems can be used in gasification processes. In the fixed-bed system, a grate supports lumps of coal through which the steam is passed. Another option is the fluidized-bed system. Finally, experimentation is occurring with entrainment systems, which use steam and oxygen to transport the coal through the reactor.

Given only the few choices described above, one can easily understand the large number of choice combinations that exist in just the first stage of a gasification process.

Upgrading. Pipeline-quality gas requires conversion of medium-Btu gas to high-Btu gas by methanation. Any of a large number of engineering choices can be made for carrying out this process.

Many alternative designs for both gasifying coal and upgrading the gas thereby produced have been tested on laboratory and pilot-plant scales. There is no consensus within the technical community in the United States on which of these options is clearly superior for large, commercial-scale plants. Until such commercial-scale plants are built and put into operation (one presently being constructed is scheduled for operation starting in the late 1980s), disagreement will remain concerning the reliability of the plants and the cost of gas produced from these varying processes.

In sum, the whole range of gasification technologies falls in the potential subsector.

Liquid Fuels from Coal. Several methods have been devised for producing synthetic liquids from coal. In all liquefaction processes hydrogen must be added to the compounds in coal. Again, the objective is to achieve the appropriate carbon-to-hydrogen ratio (bituminous coal has a ratio of one carbon atom to one hydrogen atom, while fuel oil has a ratio of one carbon atom to two hydrogen atoms).

Various pilot-scale synthetic-liquid plants using various engineering and chemical approaches have been tested. Two large synthetic-liquid plants have been constructed in South Africa using the Fischer-Tropsche process initially developed in Germany. It is generally agreed, however, that the liquids produced by the South African plants are not economically competitive with conventional crude oil.

Very substantial disagreement exists within the technical community concerning which processes will be the most reliable and efficient for commercial-scale operations. Synthetic-liquid activities thus fall in the potential subsector.

Improved Solid Fuels from Coal. A final coal-processing option involves mixing crushed coal with a solvent that has a high hydrogen content. In this process the hydrogen mixes with the coal, again raising the hydrogen-to-carbon ratio. This process also allows removal from the coal of substantial quantities of ash and sulfur and produces a solid product that is both cleaner and of substantially higher heat value. As with the gaseous and liquid options, there remains substantial disagreement within the technical community over the reliability and economic efficiency of the solvent process. Technologies for providing improved solid fuels from coal thus fall in the potential subsector.

Transportation

Raw coal is transported by railroad, barge, truck, or pipeline. Liquid and gaseous fuels produced in synthetic coal plants would be transported in the same manner as crude oil or natural gas. All these technologies fall in the available subsector.

In recent years there has been much discussion about increasing the use of slurry pipelines to transport raw coal. For a number of years such a pipeline has been used to transport coal 270 miles from a mine in Arizona to a power plant in Nevada. The success of that operation has demonstrated the availability of the slurry pipeline.

Summary

All the activities associated with mining, treatment, transportation, and use of raw coal, as well as the transportation processes for synthetic products from coal, fall in the available subsector. Some combustion technologies and all the processing technologies for converting coal into liquids, gases, or improved solid fuels fall in the potential subsector.

CHAPTER 6

Uncertain Alternatives: Nuclear Energy

In the United States no other energy source can be as directly attributed to a self-conscious, continuous government development program as nuclear power. The beginnings of nuclear energy lie in the Manhattan Project of World War II, and its use in generating electric power began in the 1950s.

The nuclear energy program was stimulated by the availability of the technology, the belief that nuclear energy could be a low-cost substitute for limited fossil energy sources, and international nuclear regulation concerns. From the vantage point of today two observations can be made about nuclear energy. First, it has developed more slowly than its advocates prophesied. For example, in 1974 the Atomic Energy Commission projected that light-water reactors (LWRs) would be providing 16 percent of total United States electric power by 1980 (INFO, 1974, p. 7). In fact, LWRs provided only slightly more than 11 percent of the nation's electricity (General Electric Company, 1981, pp. 16–17). Second, the controversy surrounding the development and use of nuclear energy has grown continually.

In concept nuclear energy can be produced by two processes: fission and fusion. In the fission process certain heavy atoms are split into two dissimilar atoms, releasing energy in the form of heat. The fusion process is conceptually just the opposite: it generates energy by fusing certain light atoms.

In the following pages are described the fission options that are receiving attention in the United States and the potential fusion options.

FISSION RESOURCE SYSTEM

Two generations of nuclear-fission technology are either available or under development in the United States in the form of

137

conventional fission reactors and breeder reactors. Conventional fission reactors are of two types: the light-water reactor (LWR) and the high-temperature gas reactor (HTGR). LWRs are in widespread commercial use in the United States, while only one commercial-scale HTGR has been built.

The breeder reactor is a second-generation fission reactor. A commercial-scale demonstration breeder is presently being designed in the United States. Experimental breeders have been in operation for several years in France, Britain, and the Soviet Union. The driving force behind the development of the breeder has been the projected scarcity of uranium fuel for light-water reactors. The breeder is attractive in concept because it uses an isotope of uranium that is in abundant supply in waste piles of uranium enrichment plants.

Resource Base

The complexity of every component of the nuclear resource system is immediately signaled in the discussion of uranium resources. The isotope U-235 is used to fuel the light-water reactor. Uranium occurs in nature as a compound. On the average one ton of uranium-bearing ore contains 0.024 to 0.030 pound of U-235 (Science and Public Policy Program, 1975, p. 6-3). Estimates of the quantity of uranium and, therefore, U-235 available in the United States vary widely and are based on a complex method of calculation. Uranium resources and reserves are expressed in terms of the quantity of yellowcake (U_3O_8), the product of first-stage milling of uranium-bearing ore, that is available at a specific dollar-per-pound level.

In periods of inflation it is necessary to recalculate constantly the effects of that inflation on every step of the finding, producing, and milling processes. In addition, unlike coal, uranium does not occur in continuous, clearly identifiable beds. Both the form in which natural uranium occurs and the system for calculating reserves contribute to the uncertainties surrounding supply.

The uncertainty about the quantity of natural uranium added to the very substantial uncertainty about the future demand for uranium fuel provides a classic example of an intractable problem. What is clear, however, is that, like oil and natural gas, uranium is a limited resource. Most of the known uranium resources in the United States are in New Mexico, Wyoming, Utah, Colorado, and Arizona. Substantial quantities of those resources are on public lands and Indian reservations.

Light-Water Reactor (LWR) System

Production of electricity by the light-water reactor system involves ten distinct activities: (1) exploration, (2) mining and reclamation, (3) milling, (4) production of uranium hexafluoride (UF_6), (5) enrichment, (6) fuel fabrication, (7) electricity generation, (8) fuel reprocessing, (9) radioactive waste management, and (10) transportation. These steps are referred to as the LWR fuel cycle (see figure 6.1).

Exploration

Exploration for uranium normally has three phases. In the first phase a survey is conducted to identify host rock, usually sandstone. The second phase involves a more detailed survey, including surface and subsurface mapping and sampling, as well as the use of various instruments to detect radioactivity. The third phase normally involves drilling into suspected deposits. Drilling permits two kinds of assessments: measurement with down-hole instruments and geochemical analyses of the materials brought to the surface. Exploration techniques and activities are well established, although improvements are being made, and are in the available subsector.

Mining and Reclamation

Uranium is mined both in open pits and in underground mines. Mining techniques are similar to those used for other minerals. The major distinction is the requirement for special ventilation systems in underground uranium mines. Uranium creates radon, a radioactive gas, exposure to which has been correlated with the occurrence of lung cancer in miners. The techniques for managing radon as well as for reclamation appear to be well within the state of the art, and all phases of mining fall within the available subsector.

Milling

As mentioned above, uranium ore is milled to produce a compound called yellowcake (U_3O_8). Approximately 500 tons of

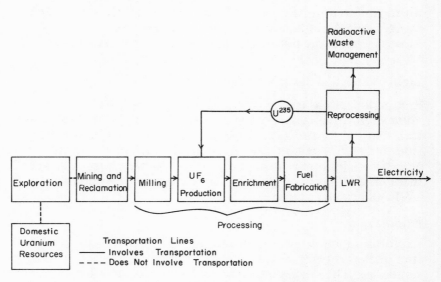

Fig 6.1. Light-water reactor fuel cycle. From Science and Public Policy Program, University of Oklahoma, 1975, p. 6-4.

ore are necessary to produce one ton of yellowcake. The milling process is a state-of-the-art available activity.

Production of Uranium Hexafluoride (UF₆)

The next step in the production of nuclear fuel is to convert the uranium in yellowcake into a gaseous compound, uranium hexafluoride (UF_6), which is used in the enrichment stage. Two processes for producing UF_6 are commercially available, have been used for many years, and therefore fall in the available subsector.

Enrichment

Enrichment is the process by which the percentage of the isotope U-235 is increased in nuclear fuel. Light-water reactors require a fuel that is approximately 3 percent U-235. In the United States at present there are one available and two proposed enrichment technologies.

The technology currently used is gaseous diffusion. In a gaseous-diffusion plant UF_6 is pumped through a large complex of pipes and filters until the desired concentration of U-235 is achieved. Existing gaseous-diffusion plants are viewed as inefficient and out of date, with capacities inadequate to meet any significant increase in the demand for enriched uranium.

In anticipation of the need to build second-generation enrichment capabilities, two additional enrichment technologies have been investigated: the ultracentrifuge process and the laser enrichment process. The ultracentrifuge process appears to be preferred.

In sum, the gaseous diffusion process falls in the available subsector, and the ultracentrifuge and laser-enrichment processes fall in the potential subsector.

Fuel Fabrication

At the fuel-fabrication stage the enriched uranium is converted into pellets, which are encased in long metal tubes called "cladding." The tubes are used in making up the fuel assembly for light-water reactors. The fabrication process is well understood and falls in the available subsector.

Electricity Generation

Light-water reactors in commercial use in the United States are of two basic designs: (1) boiling-water reactors (see figure 6.2) and (2) pressurized-water reactors. Both are similar to fossil-fuel-fired power plants except that the nuclear steam cycle replaces the conventional fuel boiler, and the nuclear fuel core replaces the fossil-fuel supply. The heat energy comes from fissioning, or splitting, the U-235 atom.

Persistent questions are raised about the design and operation of the LWR plants. They are very complex and involve the integration of large numbers of components. As in any other complex facility, the failure of individual components can cause a plant breakdown. Unreliable performance can be costly to the utilities and their customers. Another concern is the possibility of a breakdown in which the core of the reactor might overheat and melt, with the result that superheated steam would rupture the containment system and release radioactive materials into the atmosphere. Still another concern is that some external event, such as an earthquake, might cause loss of control of a reactor.

141

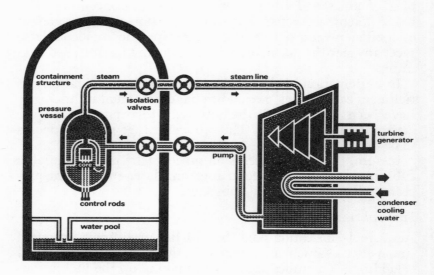

Fig. 6.2. Boiling-water version of light-water reactor. From Atomic Industrial Forum, Inc., 1974.

After many years of commercial operation, LWRs are clearly in the available subsector. It is important to note, however, that continuing technical questions arise about the ultimate reliability and safety of these complex systems. Light-water reactors are ringed with more uncertainties about performance than is the norm for technologies in the available subsector. These uncertainties, when added to the very substantial uncertainty about the long-term effects of radioactivity on human and other biological systems, complicate decision-making processes involving light-water reactors.

Fuel Reprocessing

During the LWR's early developmental period it was anticipated that used fuel would be reprocessed to recover unused U-235, as well as the plutonium created in the fission process. In 1966 the first commercial fuel-reprocessing plant began operation. In the early 1970s, however, the plant was shut down for modifications and has not been reopened. The major barriers to commercial reprocessing plants have been concerns about safety and costs. Years of experience with reprocessing military nuclear materials, however, indicates that nuclear fuel reprocessing should be classified in the available subsector.

Radioactive Waste Management

The objective of waste management is to ensure that nuclear materials do not enter the environment until their radioactivity falls below harmful levels. Certain types of highly radioactive wastes must be isolated from the environment for hundreds of years. In the absence of reprocessing operations, radioactive waste management has become increasingly important. At present large quantities of irradiated fuel are being held at the sites of the various nuclear generating plants.

Radioactive wastes are classified as either high-level or other-than-high-level waste. It was initially expected that reprocessing would separate these two categories of waste. High-level waste was to be stored in permanent or long-term facilities. Other-than-high-level waste was to be buried in shallow trenches with the expectation that it would remain there permanently.

Most of the plans for developing permanent storage facilities involved burial deep in salt or rock formations that had appropriate geologic and chemical characteristics to ensure that the waste was permanently excluded from the environment.

At present there is disagreement in the scientific-technical community on waste storage. The disagreement has the effect of ensuring local opposition at nearly any proposed permanent-storage site. Nuclear waste management thus falls in the potential subsector.

Transportation of Nuclear Materials

Transportation occurs between most of the steps in the nuclear fuel cycle: between mining and milling, between milling and UF_6 production, between UF_6 production and enrichment, between enrichment and fuel fabrication, between fuel fabrication and the power plant, between the power plant and the reprocessing facility, and between the reprocessing facility and the waste-management facility.

The nuclear transportation system must be designed to do three things: (1) protect the general public and workers from radiation, (2) ensure no release of radiation in case of accident, and (3) ensure the security of the materials.

143

The federal government has established detailed regulations covering nuclear transportation. Although the standards periodically come under criticism, transportation is well understood and falls in the available subsector.

Summary

The activities comprising the operation of a light-water reactor are called the LWR fuel cycle. The term reflects the interdependence and intricate intertwining of the activities involved in nuclear power generation. The technologies for carrying out all the steps through the generation of electricity fall in the available subsector. Only in the enrichment process did we identify alternative technologies that fall in the potential subsector. Note, however, that radioactive waste management is clearly in the potential subsector.

The absence of waste-management activities in the available subsector is a major problem for the LWR system. The ever-growing accumulation of radioactive waste is a persistent issue.

High-Temperature Gas Reactor (HTGR) System

Although one 330-megawatt HTGR is in operation in the United States, many members of the technical community question the adequacy of the reactor's performance, specifically its high operating cost in comparison to that of LWRs. As noted earlier, the development of the nuclear energy system has been heavily dependent upon funding and direction by the federal government. At an early stage the government decided to support the development of the light-water reactor in a substantial way. On the other hand, the HTGR received only very limited federal support and proved unsuccessful as a commercial venture of the General Atomic Corporation.

The HTGR differs from the light-water reactor in two fundamental ways. First, it uses helium, a gas, as the coolant and the medium for transferring heat. Second, it uses thorium-232 (Th-232) as fuel.

Thorium, one of the basic elements, occurs naturally in various chemical forms. Since the demand for thorium has been very small, thorium resources are not well identified. There is, however, general agreement that sufficient thorium

144

exists to supply the needs of large-scale use of HTGRs for the foreseeable future.

The HTGR system comprises seven steps: (1) exploration, (2) mining, (3) processing and fuel fabrication, (4) HTGR energy production, (5) reprocessing, (6) radioactive waste management, and (7) transportation (see figure 6.3).

Exploration

The methods used for uranium exploration are also used for thorium exploration. Exploration activities fall in the available subsector.

Mining

Thorium mining uses conventional mining methods. Thorium-mining technologies are, therefore, in the available subsector.

Processing and Fuel Fabrication

Although thorium processing and the fabrication of HTGR fuel appear to be well understood, no commercial fuel-fabrication plants have been built, and there is still work to be done in developing the technology for commercial production.

During its start-up period HTGR fuel requires the construction of two types of microspheres, one containing U-235 and the other containing a combination of thorium and U-233. The U-233 is created in normal reactor operations when the Th-232 is converted by the reactor process into the isotope U-233 (note once again the pattern of interdependence in the processes of the nuclear fuel cycles). The two types of microspheres then are loaded into a graphite block to make the fuel element.

It appears appropriate to place HTGR fuel processing and fabrication in the potential subsector. There is, however, very little doubt within the scientific-technical community about the nation's capacity to make HTGR fuel available on a commercial scale.

HTGR Energy Production

In the HTGR the heat from the reactor core is transferred by helium gas to a water-circulating heat exchanger, creating steam (see figure 6.4). The steam drives the turbine, which

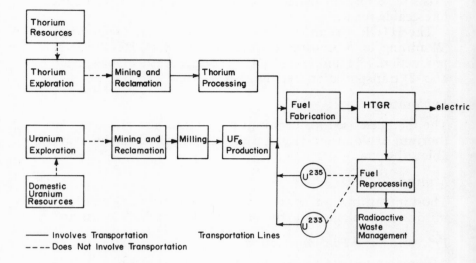

Fig. 6.3.   High-temperature gas reactor fuel cycle. From Science and Public Policy Program, University of Oklahoma, 1975, p. 6-43.

in turn drives the generator, as in other steam electric power plants. Most experts agree that in concept the HTGR has a number of features that make the chances of an accident in which radioactive materials are released somewhat less than is the case with the LWR. On the other hand, the very limited commercial-scale experience with the HTGR means that there is no large data base to confirm the theoretical safety advantages of the HTGR. The existence of one commercial-scale HTGR has not been sufficient to demonstrate its availability; therefore, the HTGR falls in the potential subsector.

Reprocessing

The reprocessing stage of the HTGR fuel cycle is an essential stage. The purpose of reprocessing is to recover unused U-235 and Th-232, as well as capture the U-233 that is created. The reprocessing technology is not completely understood, and no commercial reprocessing facility has been built. Therefore, reprocessing must be placed in the potential subsector.

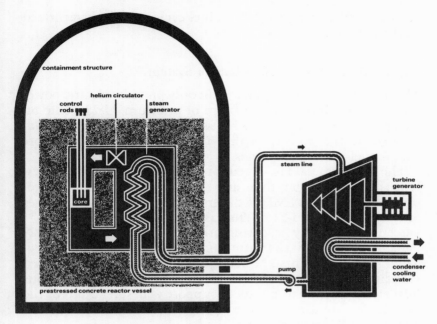

Fig. 6.4. High-temperature gas-cooled reactor. From Atomic Industrial Forum, Inc., 1974.

Radioactive Waste Management

All the problems encountered in LWR radioactive waste management are found in HTGR waste. Thus HTGR radioactive waste management is in the potential subsector.

Transportation

Much the same transportation procedures and regulations of the LWR fuel cycle would be appropriate for the HTGR fuel cycle. Transportation thus falls in the available subsector.

Summary

A review of the high-temperature gas reactor fuel cycle indicates that the specific steps of processing, the reactor, the steps of fuel reprocessing, and radioactive waste management are substantially uncertain and thus fall into the potential subsector. Therefore, the HTGR fuel cycle must be categorized in the potential subsector.

Liquid-Metal, Fast-Breeder-Reactor System

The fast breeder nuclear reactor not only provides electric power but converts an abundant isotope of uranium, U-238, into an isotope of plutonium, Pu-239 (see figure 6.5). The great advantage of the breeder is that it creates, or "breeds," more usable fuel than it consumes.

The liquid-metal fast-breeder-reactor (LMFBR) resource system is more complex than any other energy system. The resource base for the LMFBR differs from the LWR, which uses only the isotope U-235. That isotope constitutes only 0.71 percent of naturally occurring uranium. The remaining 99.29 percent of natural uranium consists of the isotope U-238. The LMFBR uses this abundant isotope. As previously noted sufficient fuel is available in existing waste piles at gaseous diffusion enrichment plants to support widespread use of the breeder for hundreds of years. There is no resource-base constraint on the development of the breeder.

The following brief summary of the LMFBR fuel cycle quickly points up its second-generation character and its dependence on the LWR fuel system.

Fuel Fabrication

Initially the fuel for the LMFBR will come from two of the activities in the LWR system. LMFBR fuel will consist of Pu-239, which will come initially from the reprocessed used fuel of LWR reactors and later from the reprocessed LMFBR fuel. To obtain Pu-239, reprocessing must be continuous. The other source of fuel for the LMFBR is U-238, obtained from existing wastes from gaseous-diffusion enrichment plants.

At present no commercial LMFBR fuel-fabrication plants exist. A number of existing plants are capable, with modification, of performing all or some of the steps necessary for fuel fabrication. There is no doubt in the scientific-technical community that fuel-fabrication plants can be built and operated successfully. Nonetheless, fuel fabrication remains a potential activity.

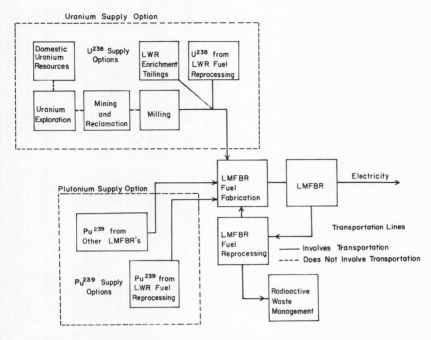

Fig. 6.5. Liquid-metal fast-breeder-reactor fuel cycle. From Science and Public Policy Program, University of Oklahoma, 1975, p. 6-60.

Reactor and Power-Generation Systems

In 1970, Congress authorized a major demonstration project for a commercial-scale liquid-metal fast-breeder reactor on the Clinch River, in Tennessee. By 1983, however, no construction had been undertaken. Breeder reactors are also being developed in a number of other countries. The French, for example, have tested them and appear committed to rapid development and use of the breeder as a way to overcome shortages of other energy resources.

Various possible designs exist for breeder reactors. The United States has put primary emphasis on one that uses the metal liquid sodium to transfer the heat from the reactor core in two phases to a point at which the sodium heats water and con-

149

verts it into steam (see figure 6.6). The liquid-metal fast-breeder reactor falls in the potential subsector.

Fuel Reprocessing

In the absence of operating breeder reactors there is not, of course, any reprocessing of LMFBR fuel in the United States. Reprocessing would be similar to that for used fuel from the LWR, but the high proportion of plutonium 239 poses a substantial number of presently unresolved problems.

One major problem is the need to account for and control large quantities of plutonium, which is a particularly important source of material for nuclear weapons. Thus control of reprocessed Pu-239 is necessary to protect against the proliferation of nuclear weapons. LMFBR fuel reprocessing is in the potential subsector.

Transportation

The transportation of large quantities of plutonium poses problems that are presently unresolved. Again, the goal is to assure that no plutonium is diverted for nuclear-weapons development. Transportation of plutonium falls in the potential subsector.

Summary

The only component of the LMFBR system that falls in the available category is the U-238 from waste piles at gaseous-diffusion plants. All other components of the breeder system must be placed in the potential subsector.

FUSION RESOURCE SYSTEM

Late in 1952, following the first hydrogen-bomb test, the Atomic Energy Commission established a project to investigate the potential of fusion as a means of generating electricity. Continuing interest in fusion has resulted from what many experts believe to be its attractive fuel and environmental characteristics. Fusion reactors are expected to be more environmentally acceptable than fission reactors because their radioactive inventories are less hazardous and have no long-lived components.

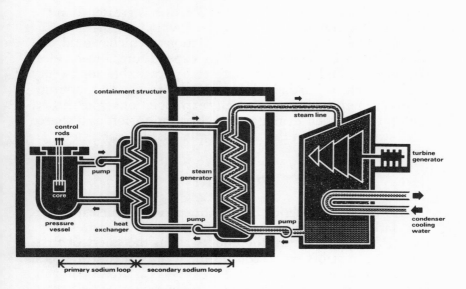

Fig. 6.6 Liquid-metal fast-breeder reactor. From Atomic Industrial Forum, Inc., 1974.

It is expected that the first fusion reactors will use two heavy isotopes of hydrogen: deuterium and tritium. Deuterium exists naturally in seawater. Tritium does not occur naturally but is expected to be produced from lithium in the normal operation of a fusion reactor. Lithium is a relatively plentiful natural element.

Two concepts for fusion reactors are being investigated: magnetic confinement and laser implosion. In the magnetic-confinement process hydrogen isotopes existing in a gas (plasma) are contained within a magnetic field. The magnetic field accelerates the isotopes to high velocities, and when the isotopes collide, fusion occurs. In the laser-implosion approach, concentrated light from lasers is used to compress and heat a pellet of deuterium and tritium, causing fusion. At present the magnetic-confinement process appears the more attractive.

Both approaches to the development of a fusion reactor are at very early stages of development. To date it has not been possible to carry out a controlled fusion reaction in which the amount of energy produced by the reaction is greater than the amount of energy used to trigger the reaction.

Under the most optimistic predictions, which would call for a major federal effort with large expenditures, fusion electric power could not be commercially available for twenty years or longer. Further, the purported energy and environmental benefits of fusion are really extrapolations from theory, and have not been demonstrated. In sum, fusion as a source of electric power is in the theoretically possible subsector.

SUMMARY

During the 1973–80 period uncertainty pervaded nuclear power. The failure to develop fuel-reprocessing and waste-storage capabilities plus continuing doubts about the safety of first-generation reactors raised continuing technical controversies. Because of the complex technology involved, the second-generation breeder reactor remained in the potential subsector. Finally, fusion remained a theoretically possible nuclear option, that is, it had yet to be proven even at a laboratory scale.

CHAPTER 7

Uncertain Alternatives: Hydroelectric Power, Oil Shale, Tar Sands, Geothermal Energy, Organic Wastes, Solar Energy, Electricity, Energy Conservation

The preceding two chapters have surveyed the four primary sources of energy in commercial use during the period from 1973 to 1980. This chapter surveys the other primary source in commercial use, hydroelectric power, plus electricity and the range of newly proposed sources of energy.

HYDROELECTRIC POWER

The first use of water power for central-station electricity generation occurred in the 1880s. In 1940, 30 percent of the installed electric-generation capacity in the United States was hydroelectric (Science and Public Policy Program, 1975, p. 9–1). By 1973 it had fallen to 15 percent, and by 1981 to 11 percent (General Electric Company, 1981, p. 14). The decline in the role of hydroelectric power resulted from three factors: limited damsites, the high capital costs of hydroelectric facilities, and the growing availability of lower-cost alternatives.

Hydroelectric power has two distinctively attractive characteristics: It is a renewable-energy source, and most hydroelectric power plants can be easily adjusted to respond to the demand for electricity by controlling the amount of water that flows past the turbine.

Water is a hydroelectric resource under conditions in which an adequate quantity or flow rate occurs together with a suitable difference in elevation between the surface of the water-storage facility and the outlet of the turbine discharge (see figure 7.1). This elevation difference is called the head. Since hydroelectric resources are renewable, they are usually calcu-

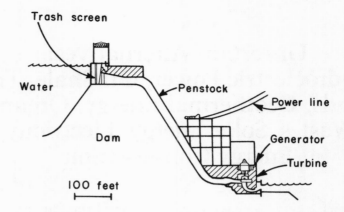

Fig. 7.1. Components of a hydroelectric power system. From Creager and Justin, 1950, p. 193.

lated as installed capacity for producing power, rather than as fixed quantities such as those calculated for fossil fuels.

Currently installed hydroelectric plants utilize many of the most attractive damsites in the United States. That fact, in combination with legislation that protects wild and scenic rivers (Wild and Scenic Rivers Act of 1968) and growing controversy over proposed new dam-construction projects, severely limits the possibility of significant increases in electricity from hydroelectric facilities.

At present most hydroelectric power comes from high dams. In recent years, however, two alternatives have received increasing attention: a return to construction of the low-head generating facilities that were common earlier in the century and the use of tidal power.

A final point should be noted with regard to hydroelectric resources. More than 40 percent of the installed hydroelectric capacity in the United States is owned by the federal government. Further, most of the attractive potential sites for hydroelectric facilities are under federal control, especially those in the western states and Alaska (Science and Public Policy Program, 1975, p. 9-3). Federal resource and land-management policies will thus play a significant role in the future development of hydroelectric energy.

The major elements in a conventional hydroelectric resource

system consist of (1) the water source, (2) the storage reservoir, (3) the transport system, and (4) the turbine-generator complex.

Water pressure for generating hydroelectric power may exist as a naturally flowing stream, but an adequate head is most often achieved by building a dam from which the water is released through a pipe and transported to a turbine, which in turn drives the generator. Dams are normally built to achieve multiple objectives, such as maintaining an adequate head for power generation, providing water storage, serving as a flood-control facility, providing a water supply, and providing recreation.

Dams are classified as either low or high. Low dams range up to about 100 feet in elevation and are normally constructed on rivers with relatively constant water flow. In low dams the running water continually turns the turbines as part of what the industry calls the base-load capacity. High dams range from about 100 to about 1,000 feet in elevation and are capable of storing great quantities of water. High dams are frequently built in mountainous areas on rivers with seasonal flows from melting snow. The flow of water from high dams can be controlled to provide extra power during periods of peak electricity demand. Hydroelectric facilities normally fall in the state-of-the-art category and are therefore in the available subsector.

One hydroelectric option that has received substantial attention in recent years is the pumped-storage facility (figure 7.2). Such a facility has two reservoirs, or pools, one at a higher elevation, the other at a lower elevation. During periods of high electricity demand, or peak periods, water from the higher reservoir is released through a turbine to the lower reservoir. When electricity demand is low, electricity from an external source is used to pump water from the lower reservoir back to the higher reservoir.

The pumped-storage system has been especially attractive in conjunction with a nuclear power plant because it allows the nuclear plant to run at a steady rate, using excess electricity available during low-demand periods to pump water from the lower to the higher reservoir. Since the energy costs for nuclear plants are relatively constant whether the plant is generating larger or smaller amounts of electricity, this system is believed to be economically attractive. Pumped storage is an available technology.

DURING LIGHT POWER LOAD
Pumping Cycle

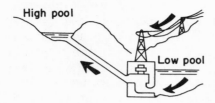

DURING PEAK POWER LOAD
Generating Cycle

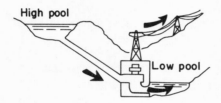

Fig. 7.2. Pumped-storage operation. From California Resources Agency, 1974, p. 13.

Tidal-Power System

The tidal-power system utilizes the gravitational pull of the sun and the moon on the ocean to generate electricity. The gravitational pull of the sun and the moon causes a bulge, or wave, of water to move across the oceans as the earth rotates. In locations where a bay partly encloses the tidal bulge, or wave, the difference between low tide and high tide may be as much as 50 feet. A few tidal-power facilities have been built and put into experimental operation; the largest facility is in France. At the present time, however, the economics of tidal-power facilities are not generally attractive. Tidal power is in the potential subsector.

Summary

All the elements of hydroelectric power except tidal power fall into the available subsector.

OIL SHALE

Oil shale is sedimentary rock containing kerogen, a solid organic material. When oil shale is heated, kerogen is decomposed into both liquid and gas. Liquid kerogen (shale oil) can be upgraded to a synthetic crude oil with the addition of hydrogen.

Oil shale has been used intermittently as an energy resource in various regions of the world for about 150 years. In the United States, before the Civil War consideration was given to developing a shale-oil industry in Appalachia, but with the discovery of oil in Pennsylvania in 1859 a cheaper and more accessible energy source became available (Science and Public Policy Program, 1975, p. 2-1). At present there is no commercial oil-shale production in the United States.

Since the embargo interest in oil shale has substantially increased. One of the goals of the federal Synthetic Fuels Corporation, established in 1980, was to assure that a major, accelerated effort would be made to develop oil-shale resources in the United States.

Resource Base

It is estimated that about 400 billion barrels of shale oil exist in high-grade shale deposits in northwestern Colorado, northeastern Utah, and southwestern Wyoming (figure 7.3; Science and Public Policy Program, 1975, p. 2-4). High-grade shale is normally defined as a deposit that averages over 25 gallons of oil per ton of shale.

Since 1973 interest in these very large oil-shale resources has grown steadily. The federal government owns 80 percent of the high-grade oil shale (Science and Public Policy Program, 1975, p. 2-9). Federal land-management policies are thus a critical element in any effort to develop a shale-oil industry.

Oil-Shale Activities

The development of oil-shale resources involves six major activities: (1) exploration, (2) mining, (3) preparation, (4) processing, (5) reclamation, and (6) transportation. The development sequence may proceed in one of two directions after the

157

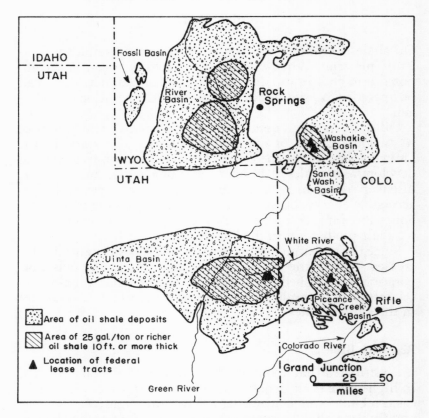

Fig. 7.3. Oil-shale areas in Colorado, Utah, and Wyoming. From U.S. Department of the Interior, 1973, 1:II-3.

exploration stage. The oil-bearing shale may be mined and then processed on the surface, or it may be processed underground (in situ). Although reclamation is a corollary of mining, it is here discussed following the processing stage because the by-products of processing pose the primary reclamation problem.

Exploration

Exploration for oil shale involves the same general activities as those for coal. Exploration falls in the available subsector.

158

Mining

Oil shale can be extracted by either surface or underground mining. Oil-shale and coal mines differ in three major ways:

1. The size of the mine needed to support a commercial-scale oil-shale facility is huge. Expectations are that each commercial oil-shale mine will need to produce approximately 25 million tons of shale a year. Each mine then, will be as large as the largest coal mines operating in the United States.

2. Oil shale is substantially harder than coal.

3. Commercially attractive shale is laid down in very thick zones.

Surface Mining. It is anticipated that surface oil-shale mining will be more like limestone quarrying or open-pit copper mining than coal mining (figure 7.4). A surface mine will likely have several working benches, and the equipment used to load, and transport the shale at the mine will be huge. At present there are neither commercial scale nor prototype surface oil-shale mines. the equipment used to mine shale will be similar to that used in other surface or open-pit mines. The only questions concerning equipment have to do with the requirements for moving massive quantities of shale. Surface mining for shale thus falls in the available subsector.

Underground Mining. Although there are as yet no commercial-scale underground shale mines, prototypes and pilot-scale mines have been opened. The tunnels or shafts into underground shale mines are much larger than those normally opened in coal mines. Large ore-carrying trucks are used underground. To date underground mining has used conventional drilling, blasting, and loading techniques. With commercial operations advanced cutting machines will likely be developed to increase the speed and efficiency of shale mining and loading.

Underground shale mines are expected to use the room-and-pillar system (see chapter 5). The rooms will be large, perhaps 60-foot cubes, and mining will be carried on at more than one level.

Oil-shale mining falls in the available subsector.

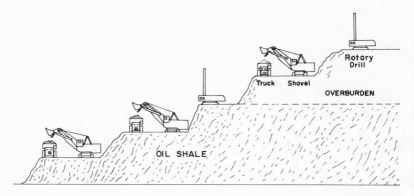

Fig. 7.4. Hypothetical oil-shale surface mine. From National Petroleum Council, 1972, p. 51.

Within and Near-Mine Transportation

Mine transportation will consist of mine-to-crusher (preparation) and crusher-to-processor links. Both links may use either large trucks or conveyors. The first link will likely use trucks, and the second link a conveyor. Given the anticipated huge size of oil-shale mines, it is expected that the surface processing plants will be mine-mouth operations, since every effort will be made to move the shale the shortest distance possible. Shale-transportation activities fall in the available subsector.

Preparation

When oil shale is mined, it tends to break into large chunks each weighing perhaps several tons. Since processing plants require smaller-sized material, crushing and sizing are required. Some appreciation of the task involved may be gained by noting that in at least one of the processes the shale must be broken down into pieces approximately three inches in diameter.

After crushing, the shale will likely be routed to a storage facility as a protection against interruption of supply. Crushing technology falls in the available subsector.

Processing

The processing of oil shale takes place in two stages. In the first stage the shale is heated to separate gaseous and liquid hydrocarbons by a form of pyrolysis reaction called retorting (pyrolysis is the heating of organic material in an atmosphere that does not allow complete oxidation). In the second stage the liquid produced is upgraded for transportation or use. Many processes are available for retorting, which can be carried out either on the surface after mining or underground (in situ).

Surface Retorting. Several types of surface retorts have been designed and tested on either experimental or pilot-plant scale. The range of options is large. Some designs introduce the crushed shale at the top of the retort (figure 7.5), while others introduce it at the bottom. Similarly, some retorts obtain their heat from combustion inside the retort, while others use heat introduced from some external unit. Retorts are available in a wide range of designs, but at present there is no agreement in the technical community about which design will produce shale oil at the lowest cost. Surface retorts thus fall in the potential subsector.

Underground (In Situ) Retorting. An alternative to surface retorting is underground, or in situ, retorting. By the underground process the shale is fractured in a cavity underground, and heat is introduced to cause pyrolysis, after which the oil is collected and withdrawn (figure 7.6).

As in surface retorting, alternative engineering approaches have been devised for in situ retorting. One approach, called horizontal sweep, involves advancing a pyrolysis zone or wall of heat horizontally from a set of injection wells to a set of extraction wells. In another approach, called mine and collapse, a pyrolysis zone or wall of heat is advanced vertically in a large underground version of an internally heated retort. In situ retorting is a potential activity.

Upgrading

Whatever retorting process is used, processed shale oil must be upgraded. The upgrading process is similar to the first stages in a conventional crude-oil refinery. It normally involves removing sulfur and nitrogen from the oil, separating the lighter hydrocarbons from the heavier ones in the oil, and then treating the lighter hydrocarbons with hydrogen to make them higher

161

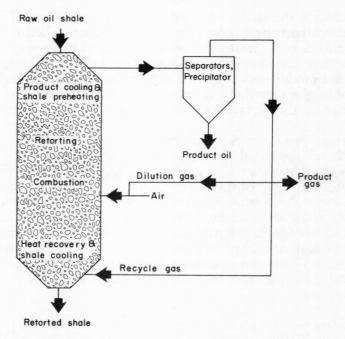

Fig. 7.5. Gas-combustion process. From U.S. Bureau of Mines, 1950.

in quality and more liquid. The upgrading stage of shale-oil processing falls in the available subsector.

Reclamation

Reclamation requires dealing with a wide range of materials produced in the mining operation, as well as large quantities of spent shale from which the hydrocarbons have been removed.

Spent shale is less dense and therefore larger in volume than raw oil shale. As mentioned earlier, the quantity of shale needed to support a commercial facility may be as much as 25 million tons a year. This makes the disposal and reclamation of spent shale a very serious problem.

Many approaches to reclamation have been proposed and investigated on a limited scale. The economically most attractive method appears to be disposal in a naturally occurring deep

162

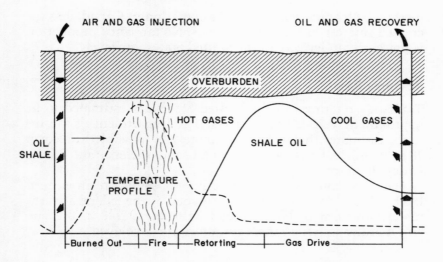

Fig. 7.6. In situ retorting process. From U.S. Department of the Interior, 1973, 1:I-37.

canyon, where the shale can be managed to prevent material leached out of it from polluting either surface or groundwater. Reclamation remains a potential activity.

Transportation of Finished Product

Upgraded shale oil can be moved in the same ways that crude oil is moved. If upgrading does not occur at the retorting site, the product, which is very thick, will require special handling (perhaps heating) before it can be transported. Transportation technologies fall in the available subsector.

Summary

All shale-oil processing activities fall in the available subsector except retorting and reclamation, which are potential activities.

TAR SANDS

Tar sands are deposits of porous rock or sediments that contain oils too thick to extract by conventional petroleum-recovery methods. A commercial tar-sands operation in Canada has increased interest in the United States. No tar-sands production is presently under way in the United States, however.

Resource Base

Tar-sands resources in the United States are estimated to be approximately 30 billion barrels. The Department of the Interior estimates that it may ultimately be possible to recover 10 to 16 billion barrels of oil from tar sands (Science and Public Policy Program, 1975, p. 5-1).

The most commercially attractive tar sands are in eastern Utah. Most of those resources are owned by the federal government. Government land policies will play a role in the rate and manner in which these resources are developed.

Tar-Sands Activities

The sequence of activities associated with producing tar sands is as follows: (1) exploration, (2) extraction, (3) refining, and (4) transportation.

Exploration

Exploration for tar sands has been largely confined to visual observation of surface outcrops or tar seeps. Little use has been made of the sophisticated exploration tools used by the oil industry. Should the need arise for more intensive tar-sands exploration, those tools would be appropriate. In the near term most tar-sands exploration will utilize drilling and coring to describe more precisely known deposits. Exploration technologies for tar sands fall in the available subsector.

Extraction

The hydrocarbon in tar sands may be extracted either by mining and transporting the sands to the surface for processing or by underground (in situ) techniques. Deep deposits will probably require the use of in situ techniques.

Where it is possible to surface-mine tar sands, bucket-wheel excavators or drag lines will be used, since the deposits are not highly consolidated. Once tar sands have been mined, the oil can be recovered by three processes: hot-water extraction, solvent extraction, and pyrolysis.

The Canadian operation uses hot water, steam, and sodium hydroxide to separate the tar from the sand. In this process the sand falls to the bottom of the tank, and the tar floats to the top. In the solvent-extraction process a solvent (such as naphtha) is mixed with the tar sands, and the tar-solvent combination is then drained off and separated. In the pyrolysis method the sands are combusted in a vessel called a coker, and the hydrocarbons are recaptured as either liquids or gases.

Two methods have been suggested for in situ recovery of hydrocarbons from tar sands. The first method involves application of heat in various forms to the sands while they are in the ground. The other method involves use of emulsifiers or solvents to dissolve the tar from the sands. These in situ processes are very similar to some of the enhanced-oil-recovery techniques.

The final step in the extraction of oil from tar sands is upgrading. Upgrading is accomplished by either or both of two methods: the thermal-breakdown and the hydrogen process. In the thermal-breakdown process the higher fractions of the tar are driven off, and the remaining part of the hydrocarbon is left in the form of coke. The hydrogen process involves changing the carbon-to-hydrogen ratio of the molecules.

Refining

The processes for refining the synthetic crude from tar sands are the same as those for crude oil.

Transportation

The methods of transporting oil produced from tar sands would be similar to the methods used for transporting crude oil.

Summary

All the activities associated with producing tar sands fall in the available subsector except those using in situ processes, which fall in the potential subsector.

GEOTHERMAL ENERGY

Geothermal steam resources were first used to generate electricity at Larderello, Italy, in 1904. One commercial-scale electric-generation facility is operating in the United States at The Geysers, north of Santa Rosa, California. The Geysers facility has a capacity of approximately 1,000 megawatts.

In addition to the facility at The Geysers geothermal steam is used for space-heating purposes at several locations. These range from distribution systems that heat homes and businesses in Klamath Falls, Oregon, and Boise, Idaho, to a rosebush greenhouse in New Mexico.

The potential contribution of geothermal energy is dependent upon the development of four categories of systems:

1. Vapor-dominated systems that primarily produce steam.
2. Liquid-dominated systems that produce a mixture of hot water and steam.
3. Geopressured systems that may produce pressurized water and dissolved methane from deep reservoirs.
4. Hot dry-rock systems—geological formations having very high heat content but containing no water to act as a heat-transport medium.

In the United States only vapor-dominated systems have been put into commercial use.

Resources

Geothermal reserves and resources are commonly estimated in terms of their ability to generate given quantities of electricity. Estimates of potential geothermal capacity have ranged from 1,000 to 60,000 megawatts (Science and Public Policy Program, 1975, p. 8-1). The uncertainty surrounding the potential geothermal resources is thus large.

Both vapor- and liquid-dominated reservoirs consist of a heat source overlaid by a permeable formation (aquifer) through which the groundwater circulates. The aquifer is capped by an impermeable formation that prevents water loss. Water and steam transport the heat from the rock to a well and finally to the surface. The distinction between the liquid- and the vapor-dominated reservoir lies in whether the heat is carried to the surface by water or steam. Liquid-dominated reservoirs are about twenty times more common than vapor-dominated reservoirs.

Geopressured reservoirs differ from vapor- and liquid-dominated reservoirs in their geological characteristics and in the depth from which geothermal energy would have to be produced. Normally, geopressured systems lie 5,000 to 20,000 feet below the earth's surface. To date there has been no production from geopressured reservoirs.

In hot dry-rock reservoirs no aquifer exists. Consequently, the production of energy from them would require water injection. To date there has been no production from this type of reservoir.

Geothermal Activities

Four activities are involved in producing geothermal energy: (1) exploration, (2) extraction, (3) pipeline transportation, and (4) electric power generation.

Exploration

Presently explored geothermal systems display surface discharges of hot water or steam. The usual initial exploration step is geologic mapping. In addition to mapping, it is possible to measure the temperature and discharge characteristics at the surface, make gravity and magnetic surveys, and carry out seismic surveys and electrical-conductivity tests. If the results of these initial surveys and tests appear promising, the next step in the process is to drill a hole.

Although all the survey techniques are well understood, the degree of confidence that potential developers have in them is low because little is known about the characteristics of geothermal reservoirs. Exploration technologies thus fall in the available subsector, but predrilling exploration activities are not supported by the conceptual and theoretical understanding necessary for confidence in the results of the technologies.

Extraction

Drilling and completing geothermal wells differ from conventional oil- and gas-well drilling in the following specifics:

167

1. Drilling penetration is slower because of the harder rock.
2. The equipment must be capable of handling high temperatures.
3. The mud system must have special cooling capabilities.
4. The well casing is used as the production string because of the high volume of fluids produced.
5. Air drilling is common.

Geothermal production begins when steam lines are connected to the wellhead. In vapor-dominated systems water is separated from steam, and the steam is cleaned of corrosion particles. Separated water is often brine, and present production systems reinject the brine into the ground (figure 7.7). The activities associated with producing vapor-dominated reservoirs fall in the available subsector.

In liquid-dominated reservoirs, the technology for cleaning, producing, using, and reinjecting fluids has not been demonstrated on a commercial scale and therefore falls in the potential subsector.

There has been no geothermal production from geopressured or hot dry-rock systems, and these processes fall in the theoretically possible subsector.

Transportation of Geothermal Steam

Generation facilities must be close to the production point to reduce pressure and temperature loss in pipelines between the wellhead and the facility. Transportation systems are in the available subsector.

Electric Power Generation

Power-generation methods in geothermal systems are similar to those of other steam electric systems. Geothermal systems are distinctive primarily in that they use low-pressure steam turbines to drive the generators. Their primary energy-to-electricity efficiency is about 15 percent as compared to about 35 percent in fossil-fuel-fired steam-generating plants. The technology used to generate electricity in geothermal systems falls in the available subsector.

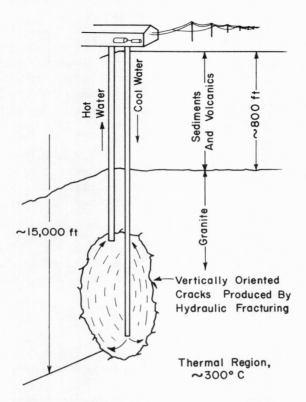

Fig. 7.7. Dry-rock geothermal energy system employing hydraulic fracturing. From Atomic Energy Commission, 1973, p. A.4-20.

Summary

All the activities associated with generating electricity with steam from vapor-dominated reservoirs fall in the available category. All other geothermal options fall in the potential or theoretically possible categories. At present questions both about technologies and about the quantity of geothermal resources available make the potential of geothermal energy highly uncertain.

169

ORGANIC WASTES

Although in Europe electricity has been generated for many years by burning of municipal wastes, no similar pattern existed in the United States at the time of the oil embargo. With the recent rise in the cost of energy, increased attention has been focused on the use of organic wastes. Experiments have been undertaken with a number of technologies for converting organic wastes into usable energy.

Although the total energy potentially available from organic wastes is only about 2 percent of the nation's needs, utilization of wastes is particularly attractive because of the additional benefit to waste management (Science and Public Policy Program, 1975, p. 2-1).

Only dry, organic solid wastes can be converted into energy. Such waste includes some categories of municipal refuse, manure, agricultural waste, logging and wood-manufacturing waste, and sewage sludge, as well as some forms of industrial waste.

Resource

Conceptually, organic wastes represent a renewable resource, and the resources and reserves are expressed as the amount available each year. Since organic wastes have a high ratio of volume to energy, it is practicable to move them only short distances. For example, manure-conversion facilities must be constructed near large-scale animal feedlots. Agricultural and wood-manufacturing wastes should be processed at the site where the wastes are produced. Urban refuse and sewage sludge offer other concentrated sources of wastes.

Organic-Wastes Activities

Organic-waste energy systems have collection and processing or burning components.

Collection

Collection activities are usually carried out with trucks or with specially designed equipment. Collection technologies for organic wastes fall in the available subsector.

Processing

Organic-waste processing converts the wastes into a usable fuel form, a liquid, gas, or electricity. The first processing stage, called beneficiation, is carried out in two steps: (1) separation of organic from inorganic wastes and (2) sizing of waste materials to an input particle size. Technologies for converting beneficiated wastes into liquid or gaseous fuels fall into three general categories: (1) hydrogenation, (2) bioconversion, and (3) pyrolysis.

Hydrogenation involves adding hydrogen to an organic molecule to achieve a higher hydrogen-to-carbon ratio. Hydrogenation introduces carbon monoxide and steam into a high-pressure, high-temperature chamber containing organic wastes and a catalyst.

Bioconversion employs microorganisms to convert organic waste into methane (natural gas). Bioconversion, in one form also known as anerobic digestion, is employed in sewage-treatment processes.

Pyrolysis involves heating organic material at atmospheric pressure in an atmosphere in which less than total combustion occurs (figure 7.8). Pyrolysis is considered technologically the most attractive process, but without modifications it produces gas and oil of lower heat values than those produced by hydrogenation and bioconversion.

A number of electricity generating plants are now burning organic wastes directly as a substitute for coal.

Summary

Collection, processing, and direct burning of organic wastes fall in the available subsector. Processes for converting organic wastes into either a liquid or a gas by hydrogenation, bioconversion, and pyrolysis fall in the potential subsector.

SOLAR ENERGY

Solar radiation provided much of the energy used in preindustrial and early industrial societies. Many buildings were constructed to take advantage of direct solar radiation. As wind,

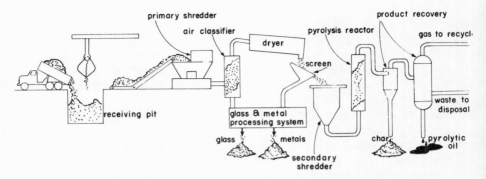

Fig. 7.8. Garrett pyrolysis system. From Garrett Research and Development Company, Inc.

solar energy was used to dry grain, pump water, and drive ships. As firewood, solar energy heated homes and buildings and provided steam for steam engines. In time more efficient energy systems replaced solar energy, which provided only a minute amount of the nation's energy in 1973. When the nation was faced with an energy shortage, efforts were made to move back to solar energy. It was seen not only as inexhaustable but also as environmentally benign.

The amount of solar radiation that reaches the surface of the earth is huge, variable, and diffuse. Given its variability, either energy storage or backup power is needed to provide energy at night or when the sun is obscured. Because of the low density of solar radiation, large land areas must be devoted to energy collection. Finally, solar energy sources tend to operate at low efficiency, and this, added to its variability and diffuseness, means that even though the energy itself is free the capital investment for collecting, storing, and transforming it is high. For these reasons most solar energy forms have not been economically competitive with other forms of energy.

Solar energy is usually divided into four categories: (1) direct radiation, (2) wind, (3) organic fuels, and (4) ocean thermal gradients (figure 7.9). In this section we discuss each of these optional solar energy forms.

172

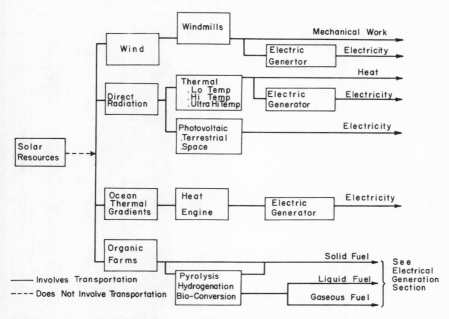

Fig. 7.9. Solar energy resource development. From Science and Public Policy Program, University of Oklahoma, 1975, p. 11-6.

Direct Solar-Radiation Technology

Solar radiation may be used directly by a photovoltaic cell, to heat an object (as a home water heater), or to heat a working fluid that transfers the heat to some ultimate receiver. There are three categories of solar heat collectors: low-temperature collectors, high-temperature concentrators, and ultrahigh-temperature concentrators.

Low-Temperature Collectors

Low-temperature solar heat collectors are in commercial use throughout the world (figure 7.10). Such collectors normally involve panels with a surface that is painted black and covered with a sheet of clear glass spaced about one-half inch above the surface. Low-temperature collectors can achieve a temperature of 225°F to 250°F under favorable conditions.

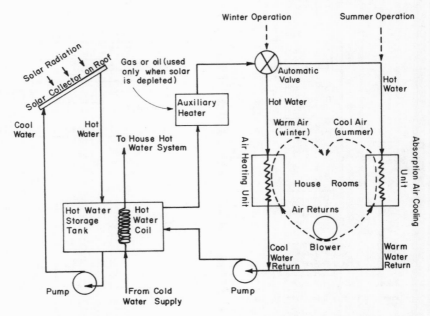

Fig. 7.10. Residential heating and cooling with solar energy. From Atomic Energy Commission, 1974, p. A.5-17.

Such collectors are commonly used for space- and water-heating purposes. Water heated in low-temperature collectors can be kept in insulated tanks to store heat during the night or when the sun is obscured.

High-Temperature Concentrators

To attain temperatures higher than 225°F to 250°F, the sun's rays must be concentrated by the use of reflecting surfaces. When water pipes are run along the focal lines of reflecting surfaces, temperatures up to 600°F are possible; therefore, high-temperature concentrators can be used to generate steam. For maximum efficiency such concentrators must track the sun. Steam from high-temperature collectors can be used to generate electricity.

Ultrahigh-Temperature Concentrators

Temperatures approaching 5,000°F can be obtained through

174

the use of multiple, precisely contoured reflectors that direct the sun's rays to a single point. These ultrahigh-temperature concentrators have the capacity to generate steam for electricity or to process heat.

Photovoltaic Cells

Photovoltaic cells convert solar radiation directly into electrical current. They were first used on an American space satellite in 1958, and they have become the major source of power for space vehicles. By general agreement photovoltaic cells have tremendous potential.

At present, except for highly specialized uses, photovoltaic cells are not economically competitive, but they are expected to become commercially competitive within a decade and are the focus of a major R&D effort.

Summary

Low-temperature solar collectors fall in the available subsector. High-temperature and ultrahigh-temperature concentrators and solar cells fall in the potential subsector.

Wind Energy

The most promising geographical locations for wind-power generation in the United States are along the coasts and through the Great Plains from Texas through the Dakotas.

Small-scale wind-driven generators designed to provide energy to single-family homes are commercially available (figure 7.11). The intermittent action of the wind, however, means that homes with wind-power generators must have backup from other sources.

There have been experiments with much larger commercial generators. Large-scale generation wind-power systems have also been investigated. These systems usually involve large-scale wind farms, which are concentrations of a large number of wind-driven generators. At a more expanded level, multiple wind farms as part of an electric power grid system have been proposed as a means of ensuring that electricity would always be available since the wind is always blowing somewhere.

In general, the technology for wind-driven generators is understood and falls in the available subsector. Nevertheless,

175

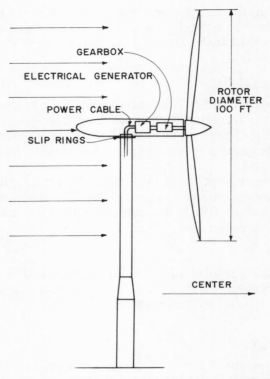

Fig. 7.11. Typical wind rotor system. From Science and Public Policy Program, University of Oklahoma, 1975, p. 11-16.

without additional development to produce greater efficiency or unless there are significant increases in the costs of other sources of energy, wind power will not be an economically competitive source of electricity on a large scale.

Organic Fuels

Organic material specifically grown to provide energy has received substantial attention in recent years despite the fact that it is a relatively inefficient energy process. Only about 1 percent of solar energy is converted into organic material in most high-yield crops. As a result, the land area required for

176

a given energy output is very high in comparison to that required by other solar-power options. In addition, there are major questions about whether land should be used to produce products for energy generation as opposed to food or, in the case of forests, building materials.

Organic material can be burned directly or converted to gaseous or liquid fuels by the same methods used for processing organic wastes. Commercially produced alcohol from such grains as corn and wheat is presently mixed with gasoline to make gasohol.

The production of biomass and its conversion into alcohol fall in the available subsector. All other liquid and gaseous biomass options fall in the potential subsector.

Ocean Thermal Gradients

The surface temperature of the oceans between the Tropic of Cancer and the Tropic of Capricorn stays remarkably constant at about 77°F. Temperatures at lower levels of the oceans vary considerably from those of the surface. This temperature difference between surface waters and waters at greater depths can be used to generate electricity in a "heat engine."

Heat exchangers are the critical technology in ocean-thermal-gradient power plants. Heat exchangers must transfer an enormous amount of heat through very thin walls while operating in a corrosive seawater environment. Although several prototype experiments have been carried out in an effort to develop this concept, many technical problems must be resolved before it will be an available option. In sum, energy from ocean thermal gradients falls in the potential subsector.

Summary

Of the components of the various solar options, low-temperature solar collectors, wind energy generators, and alcohol production fall in the available subsector. High- and ultrahigh-temperature concentrators, photovoltaic cells, ocean-thermal-gradient systems, and nonalcohol liquid and gaseous fuels from biomass fall in the potential subsector.

ELECTRICITY

Between 1960 and 1973 electric power consumption in the United States grew at an annual rate of 6 to 7 percent, nearly twice as rapidly as overall energy growth. At the time of the oil embargo the generation of electric power consumed roughly 25 percent of the nation's primary energy. By 1981 growth in electric power consumption had declined to about 3.5 percent a year, but 33 percent of the nation's primary energy was being used to generate electric power (General Electric Company, 1981, pp. 16–17). The use of electricity developed rapidly because it is clean, efficient, and versatile at the end-use point, and can be produced from all primary energy resources.

At the time of the oil embargo five primary energy resources were used to generate electricity: (1) coal, 44 percent; (2) natural gas, 18 percent; (3) oil, 18 percent; (4) hydro, 15 percent; (5) nuclear, 5 percent (General Electric Company, 1981, pp. 14–15).

Four major problems have motivated efforts to modify or find substitutes for existing electric power technologies:

1. Environmental, health, and safety problems associated with large coal fired and nuclear power plants.
2. Inefficiency of the steam electric power generation process, which converts only about 35 percent of the primary input into electricity.
3. Perceptions of oil and natural-gas scarcity and limited hydroelectric power generation sites.
4. Wide variations in the demand for electricity with the time of day and the season.

Conceptually there are five types of electric power plants plus the solar and hydroelectric options described above. They are steam (boiler-fired) power plants, combustion turbine plants, combined-cycle combustion turbine plants, fuel-cell plants, and magnetohydrodynamic (MHD) plants (figure 7.12).

Steam Power Plants

Steam power plants are the primary source of electric energy. In steam plants electricity is generated in three energy conver-

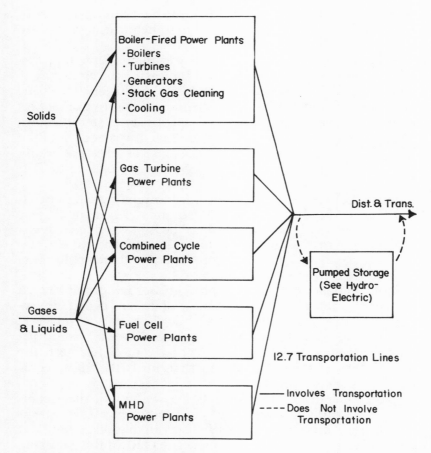

Fig. 7.12. Electrical generation system. From Science and Public Policy Program, University of Oklahoma, 1975, p. 12-2.

sion stages: (1) chemical to heat (or nuclear to heat), (2) heat to mechanical, and (3) mechanical to electrical.

A simplified description of the process is as follows: In a boiler, heat from conventionally fueled fires or from nuclear or solar sources is transferred to water to produce steam. The steam enters a turbine, where it expands to a low pressure and cools to a low temperature and drives the turbine that turns the generator. The steam discharged from the turbine is then

179

reconverted to water in a condenser, the heat being ejected into cool bodies of water or into the atmosphere through cooling towers.

Designers of fossil-fired steam power plants have been experimenting with options aimed at increasing the efficiency of the systems and reducing environmental pollution. The focus of efforts to improve both the efficiency and the environmental performance of fossil-fired power plants has been on three components: improved combustion of the fuel, improved use of the heat, and improved cleaning of the smoke or stack gas. Many alternative engineering designs have been proposed to deal with each of these problem areas.

One widely investigated option for improving combustion is called the fluidized bed (discussed under "Coal" in chapter 5). In experimental designs fluidized beds have offered advantages both in increasing the amount of heat transferred to water to create steam and in reducing the quantity of air pollutants emitted from the boiler.

An example of the effort to make greater use of the heat in the steam is represented by a number of designs called binary-cycle systems. The binary cycle attaches to the steam cycle another fluid-circulating system, which drives a second turbine. An important event in pollution control has been the development of stack-gas cleaners that purify the smoke coming from the boilers.

Fossil-fired steam power plants are well-established, commercially available systems for electricity generation. Stack-gas cleaners are also commercially available. Conventional steam plants and stack-gas cleaners thus fall in the available subsector. The various technologies for improving combustion and making greater use of the heat in the steam fall in the potential subsector.

Combustion Turbines and Combined-Cycle Power Plants

During the 1960s and early 1970s the utility industry made increasing use of combustion turbines fueled primarily by gas to generate electricity to meet peak load demands. A combustion turbine is essentially the same as a jet engine. Combustion turbines are particularly attractive for peak load power because they can be started up and shut down very rapidly. Unfortunately, turbine engines are inefficient. In an effort to improve combustion-turbine efficiency, a number of designs have been put forward for combined-cycle power plants (figure 7.13). In its simplest form such a plant uses the hot exhaust gas from the combustion turbine to convert water into steam, which is then used to drive a conventional steam cycle.

Combustion turbines as well as combined-cycle power plants fall in the available subsector. The rising cost of natural gas, however, makes turbines an unlikely answer to electricity-generation problems.

Fuel-Cell Power Plants

A fuel cell uses electrochemical reactions to produce electricity without the need for intermediate heat-mechanical cycles (figure 7.14). Theoretically the fuel cell should allow all of the energy in oil, natural gas, or coal to be converted to electricity, thereby eliminating the two-thirds energy loss associated with steam electric power plants.

Although small fuel cells are available, considerable work still needs to be done to determine whether large-scale fuel cells offer advantages in terms of economics, environmental impacts, and energy conversion efficiency. Fuel cells, therefore, fall in the potential subsector.

Magnetohydrodynamic (MHD) Power Plants

Magnetohydrodynamic (MHD) generators are theoretically attractive because they produce electrical energy directly from thermal energy, bypassing the step in conventional boilers in which heat energy is converted into mechanical energy—the step at which the steam drives the turbine and the turbine drives the generator (figure 7.15). MHD generators have the theoretical potential for converting fuels to electricity with efficiencies in the range of 50 to 60 percent.

Although work is continuing on MHD generators, there are at present no prototype plants. Therefore, they fall in the theoretically possible sector.

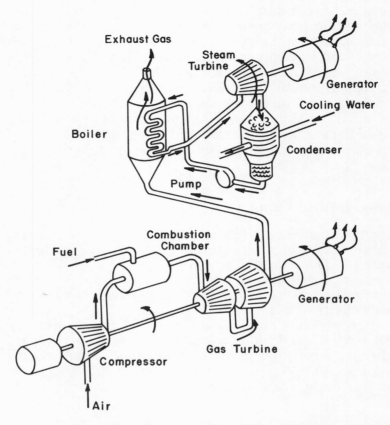

Fig. 7.13. Combined-cycle gas turbine. From Atomic Energy Commission, 1974, p. B.4-5.

Transmission and Distribution of Electricity

Normally electricity is moved from a generating plant to a user through a two-stage system. The first stage, transmission, carries the electricity at high voltage. The second stage, distribution, ultimately transmits electricity to consumers, usually at 120 to 240 volts. Between the two systems are stations or substations that convert the high voltage into the voltage used by the consumer.

Many investigations are under way aimed at increasing the

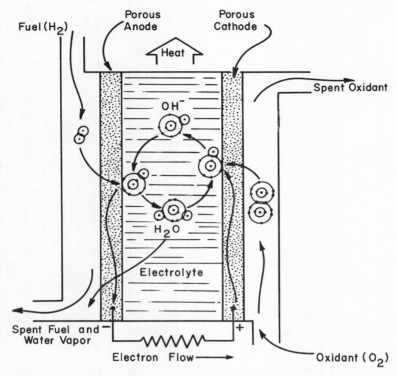

Fig. 7.14. Hydrogen-oxygen fuel cell. From Atomic Energy Commission, 1974, p. B.6-2.

efficiency of transmission lines, including studies on cryogenic, or supercooled, conductors. When certain metals are supercooled, they lose their resistance and achieve a condition known as superconductivity. Thus relatively small wires maintained in a state of superconductivity could carry very large amounts of power.

Transmission and distribution activities fall in the available subsector. Certain transmission options that are being investigated, such as cryogenics, fall in the potential subsector.

ENERGY CONSERVATION

Until 1973 energy conservation received little attention. The

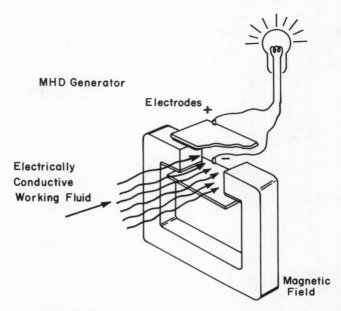

MHD Generator

Electrodes

Electrically Conductive Working Fluid

Magnetic Field

Fig. 7.15. Magnetohydrodynamic generator electrical system. From Atomic Energy Commission, 1974, p. B.10-3.

American experience with energy had been that the larger the quantities of energy produced and consumed, the cheaper the unit cost. With growing affluence and declining unit cost energy consumption increased exponentially. For example, per capita energy consumption in the United States grew approximately 70 percent between 1950 and 1970 (U.S. Department of Commerce, 1979, p. 602).

In the 1970s, Americans crossed the Rubicon so far as the unit cost of energy was concerned. After 1973 the price of energy began to escalate rapidly. As a result, conservation became a major focus of attention.

Energy conservation has been discussed under two labels: "belt tightening" and "leak plugging." "Belt tightening" refers to changes in life-style ranging from lowering the thermostat in the winter and raising it in the summer to driving at reduced speeds. "Leak plugging" means using energy more efficiently through, for example, improved insulation in buildings, more efficient factories, and more energy-efficient automobiles.

184

Energy conservation can apply to either the production or the consumption side of the energy equation. This section addresses the consumption side. It is important to note, however, that efficiency improvements in the production of energy also fall in the category of energy conservation. Therefore, energy conservation is possible in the resource systems previously discussed. For example, energy would be conserved if it were possible to save some of the two-thirds' energy loss in steam electric power plants.

The following sections focus on the three large sectors of energy consumption: (1) residential-commercial, (2) industrial, and (3) transportation. In 1973 approximately 35 percent of the energy consumed in the United States fell in the residential-commercial sector, approximately 40 percent in the industrial sector, and approximately 25 percent in the transportation sector (General Electric Company, 1981, p. 17).

A brief review of the nation's progress in energy conservation shows that between 1973 and 1975 energy consumption in the United States declined by approximately 5 percent (U.S. Department of Commerce, 1981, p. 579). From 1976 through 1979 energy consumption rose by approximately 10 percent. During 1980 and 1981 energy consumption declined by about 6 percent (Bureau of National Affairs, 1982a, p. 242). The two periods of decline followed very closely the periods in which the price of OPEC oil rose rapidly—the periods following the oil embargo and the revolution in Iran.

Residential-Commercial Sector

There are three approaches to energy conservation in the residential-commercial sector. They are (1) simple conservation measures; (2) improved insulation, building design, and construction; and (3) improved equipment efficiency.

Simple conservation practices include regulating thermostats, turning off unused lights, and drawing blinds and draperies against the sun's heat. Such options are either low- or no-capital cost options, but they generally require changes in personal behavior.

Thermal insulation is clearly the area with the most potential for energy conservation in the residential-commercial sector. In older structures conservation involves adding storm

185

windows and doors, adding caulking and weather stripping, and insulating attics. For new structures it means integrating energy-conservation criteria into the design and constuction. These options usually require an initial additional capital cost, but with rising energy prices they generally are economically viable options when calculated over the life of the building.

Nearly every appliance and heating or cooling unit offers opportunities for improved efficiency. For example, more efficient furnaces and air-conditioners require additional capital costs, but energy savings normally pay back those costs in a reasonable period of time.

Replacement of electric-resistance heating with heat pumps can be a particularly important way to conserve energy. The heat pump is essentially a refrigeration system capable of operating in reverse. Heat is pumped outdoors to provide summer cooling and pumped indoors to provide winter heating. In general, heat pumps deliver about two units of heat for each unit of electrical energy used.

All the options commonly identified with energy conservation in the residential-commercial sector fall in the available subsector.

Industrial Sector

More than half the energy consumed in the industrial sector is used for what are identified as "thermal processes," that is, direct burning of fuels or the manufacture of steam. Within the industrial sector manufacturing activities consume approximately 85 percent of the energy used, the remainder being divided approximately equally between agriculture and mining.

Two points should be made about energy conservation in the industrial sector. First, the very large number of activities for which energy is used in the industrial sector precludes any simple listing of the energy-conservation options that are available. In almost every activity, however, opportunities for energy conservation exist. Second, in the industrial sector both the capability and the inclination exists to balance the cost of energy against the cost of capital investments in modifications and procedures that save energy.

As the price of energy went up following the oil embargo, it was the industrial sector that initially responded most effectively with leak-plugging actions. This point can be clearly illustrated in the instance of an older oil refinery. The steam used in the refinery was heated by natural gas. Before the oil embargo approximately 30 percent of the steam was lost through leaks and poor insulation of steam pipes. Natural gas was so inexpensive that it was cheaper to buy the gas than to plug the leaks. As the price of natural gas rose, the refinery immediately initiated efforts to capture the lost energy.

In sum, then, a great many conservation options are available to the industrial sector that fall in the available sub-sector. There is general confidence that the industrial sector will implement energy conservation when it is economically attractive to do so, that is, when the capital cost of achieving energy conservation is lower than the operating costs associated with buying the energy fuels that are wasted. It is also doubtless the case that improved equipment designs and alternative processes exist that fall in the potential subsector.

Transportation Sector

Over one-half of the petroleum consumed in the United States is used in the transportation sector. Conservation in this sector is, therefore, critically important to the nation's level of dependence on oil imports.

For purposes of this brief discussion the transportation sector is divided into two categories: freight and passenger transportation.

Freight Transportation

Five modes of freight transport are regularly used in the United States: aircraft, trucks, waterways (barges and ships), railroads, and pipelines. The amount of energy used to move one ton of freight one mile among these five freight transportation modes varies tremendously. For example, it takes fifteen times as much energy to move one ton of freight one mile in an aircraft as it does to move the freight the same distance in a truck (Science and Public Policy Program, 1975, p. 13-36). Similarly, railroads are approximately four times more energy-efficient than trucks in moving freight, and pipelines are half again as efficient as railroads in moving liquid materials. These measures do not, of course, take into account the flexibility of trucks to go anywhere or the speed of aircraft transportation. There was a distinct movement away from railroads to trucks in the post–World War II period. Doubtless that resulted in part from the low cost of energy to fuel the trucks.

Two general categories of energy conservation opportunities are available in the freight-transportation area. One is to move larger quantities of freight by the more efficient energy modes, for example, to move more freight by rail and less by truck. The second option is to achieve greater efficiency in the use of the fuel. For example, the newer airplane engines are much more fuel-efficient than older ones.

Passenger Transportation

What has been said about freight is generally applicable to passenger transportation. Opportunities for conservation in passenger transportation split between using the most energy-efficient mode and gaining improved efficiency within each of those modes.

Clearly cars and airplanes are much less energy-efficient than trains and buses in transporting passengers. Clearly also they are much more flexible, and, of course, airplanes are much more rapid means of transportation. In general, an airplane uses roughly five times the energy required by a bus to transport a passenger one mile, and a car uses nearly twice as much energy as a bus.

Greater use of urban mass-transit systems would save a great deal of energy, as would car or van pooling. Similar efficiency improvements are possible in each mode. In the last few years the average number of miles per gallon achieved by new cars has nearly doubled over the mileage average in 1973. It appears that incremental auto-efficiency improvements will continue, saving additional energy.

In sum, most of the activities associated with energy conservation in the transportation sector fall in the available subsector.

SUMMARY

Chapters 5 to 7 have sought to provide a summary of the wide range of energy activities in the United States and to identify those options that could be taken in response to the energy shortage. One of the dilemmas faced by the various presidents and Congresses was that so many options were offered that making a choice among them was extremely difficult.

In addition, uncertainty pervaded a very large number of those choices. With regard to oil, gas, and uranium there was uncertainty about the quantity of the resource that remained to be produced in the nation. In the other resource systems, while the quantity of the resource itself was not an area of major uncertainty, the performance of the technologies and activities associated with producing and using the resource was uncertain.

It should be emphasized that the uncertainties identified reflect a lack of consensus among the scientific-technical members of the energy policy community. For any activity or technology to be classified as potential, there had to be disagreement among scientists and engineers about when the technology or the activity could be made available and what it would cost.

Policy is particularly constrained when experts disagree over the performance characteristics of activities and technologies. Policy choices are difficult enough when the participants in the policy system know who will benefit and who will suffer from particular policy choices. It becomes even more difficult to make policy when there is uncertainty about the performance of substantive activities. Performance uncertainty was precisely the situation with regard to many of the energy alternatives that were proposed during the middle and late 1970s. In sum, the energy policy participants during the 1970s had no consensus on which of the newly proposed energy options could be made available.

The lack of a consensus within the energy scientific-technical community was a major inhibitor to the rapid formulation of a national energy policy by the president and Congress. Only as these uncertainties were reduced during the latter part of the 1970s was it possible to evolve a consensus on the nation's future sources of energy. The presidential-congressional decision-making process that occurred between 1973 and the end of 1980 is the topic of the following chapters.

CHAPTER 8

The Path to Stability:
Oil, Natural Gas, Coal

The ultimate question is whether this society is willing to exercise the internal discipline to select and pursue a coherent set of policies well in advance of a threatened disaster. Western democracies have demonstrated such discipline in the past in reacting to immediate, palpable threats to survival, as in time of war. But they have had less success in harnessing their human and material resources to deal with less visible and immediate threats to their political and economic systems. When dangers appear incrementally and the day of reckoning seems far in the future, democratic political leaders have been reluctant to take decisive and perhaps unpopular action. But such action will be required to meet the energy crisis. If the nation continues to drift, it will do so in an increasingly perilous sea. —National Energy Plan, White House Release, April 29, 1977, page 25*

The confused and uncertain state of the energy issues that arrived on the presidential-congressional agenda in 1973 can hardly be exaggerated. In the seven years following the oil embargo, three presidents and four Congresses struggled continuously with those issues in an effort to formulate a national energy policy and construct an energy policy system.

The twelve resource options covered by the label "energy" shared little in the way of a common typology or nomenclature. The nation's tradition of focusing on specific fuels as the loci of policy left the various presidents and Congresses with neither a conceptual nor an organizational framework that could simplify and clarify the nature of the debate and the issues involved.

The vigorous efforts of a broad and diverse set of actors and interests to influence the nation's energy policy, in combination with a large and uncertain set of resource options, complicated decision making in the extreme. Before an energy policy could be formulated and a stable energy policy system constructed, a difficult and conflict-ridden sorting process was necessary that demanded continuing presidential-congressional attention through 1980.

Four broad issues had to be resolved before the conditions would exist for energy policy stability. First, priorities had to be established among the four national energy goals: abundance, cheapness, cleanness, and security. Second, choices had to be made about which energy sources would supply the nation's needs in the future. Third, it was necessary to establish priorities with regard to the instruments that national policy would utilize in promoting the development of the energy sources that were to achieve the nation's energy goals. Finally, a difficult sorting process had to take place to determine which actors and interests would be continuing participants in the new energy policy system. The following chapters investigate and analyze the evolution of the presidential and congressional decisions that were made through the end of 1980.

It will be recalled that stable policy systems require agreement in two basic areas. First, there must be agreement on the boundaries of the policy sector. In the area of energy the various presidents and Congresses were involved in making choices about which of the twelve resource options would receive priority. Second, there had to be agreement on which actors and interests would participate in the energy policy community. With the breakdown of the fuel policy systems a multitude of actors and interests were striving for admission to the new energy community.

During the period from 1973 through 1980 a process of issue resolution occurred at the presidential-congressional level that was to lead to a rudimentary national consensus on energy. As discussed in chapter 2, what occurred with energy was consistent with traditional patterns in the United States when major new policy is being developed.

Before the oil embargo stable energy management was provided by the five fuel policy systems. After 1973 those fuel policy systems were no longer capable of providing stable energy management. The issue-resolution process that occurred through 1980 was a transition period. During this transition compromises were evolved that established the rudimentary consensus necessary for a new stable energy policy system, one that had a high probability of being able to manage energy through a process of problem solving.

Two things are striking about the process of issue resolution discussed in these pages. First, most of the fundamental choices and compromises ultimately evolved were very similar to the early proposals of President Richard M. Nixon. Although Nixon was strikingly unsuccessful in getting the Congress to adopt his proposals, they were to become the framework for the consensus that was finally developed.

Second, that much-maligned institution, the Congress, provided the continuity and pattern of moderation necessary to produce a consensus. Although observers have repeatedly criticized the lack of decisiveness in the legislative branch with regard to energy, from hindsight that criticism appears largely unwarranted. In fact many of the more innovative initiatives in the solar energy and energy conservation areas came from the Congress, not the executive branch. Given the high levels of uncertainty associated with many of the proposed energy alternatives plus the intense advocacy manifested by differing actors, decisiveness would have been inconsistent with the American policymaking tradition.

Chapters 8 and 9 are organized around the twelve resources because those resources provided the structure for specific policy actions. The genius of the process was that the specific policy decisions in this period led to an aggregated energy system as opposed to the previously disaggregated fuel systems. Following the investigation of the actions on the twelve resources in chapters 8 and 9, chapter 10 describes the process that led to the identification of the major participants in the new energy policy community. Choices with regard to both the nation's energy goals and the policy instruments to be used were made primarily in the context of choosing among the various sources of energy and in the selection of the continuing participants in the energy policy community.

OIL

Price Deregulation

On April 5, 1979, President Jimmy Carter announced the deregulation of domestically produced oil. Carter's decision initiated a process of gradual price decontrol that started June 15, 1979, with total decontrol to occur on September 30, 1981 (U.S. President, 1979, p. 609).

Two things are striking about the decision to decontrol oil. First, by 1979 previously contending actors and interests had coalesced in support of deregulation. Even such traditional congressional opponents of deregulation as Congressman John D. Dingell had concluded that federal control of the price of oil was a failure (Congressional Quarterly, Inc., 1981, p. 118). Advocates of other energy sources, ranging from solar energy to coal, saw increased opportunities for their favorite energy sources if there should be a significant increase in the price of oil. Finally, conservation interests had concluded that oil priced at its "true" replacement value was the most effective instrument for stimulating conservation.

The decision to deregulate oil was a significant addition to a process of price deregulation that had started in 1978 with the passage of the Natural Gas Policy Act. This act followed by the deregulation of oil represented a fundamental decision concerning the relative priorities of national energy goals. With these decisions the president and Congress reluctantly chose to sacrifice cheap energy as a national goal. It finally became obvious to a majority in Congress that cheap energy was no longer possible.

Oil prices had been regulated by the federal government since August 16, 1971, when President Nixon imposed an economy-wide freeze on wages and prices (Congressional Quarterly, Inc., 1973, p. 108). The freeze was aimed at controlling inflation. Nixon's price-freeze decision was the beginning of eight years of tortuous efforts by the federal government to orchestrate a package of price incentives and sanctions for oil that became ever more complex.

The difficult problems rising from price regulation were signaled almost immediately. At the time of the 1971 price freeze, gasoline prices were high and fuel-oil prices were low, reflecting a summer market. The freeze made gasoline more profitable to produce, and refiners put more emphasis on producing gasoline than they did on producing less profitable fuel oil. The result was that the price freeze was seen as contributing to a shortage of fuel oil in the winter of 1971-72 (*Oil and Gas Journal,* 1971b, p. 87).

The 1971-72 fuel-oil shortage was only the first of many problems during the next eight years that led to widespread public disenchantment with the government's ability to regulate effectively the price of oil. It became increasingly apparent that the federal government simply was not able to manage effectively the highly complex and detailed decisions that were required to regulate oil.

Shortly before the oil embargo was imposed in 1973, the president's Cost of Living Council, which had responsibility for containing inflation, implemented a new system of regulation specially aimed at the oil industry as a part of Nixon's Phase 4 control program. This new system set up two categories of oil: "old oil" and "new oil" (Congressional Quarterly, Inc., 1981, p. 31). New oil was that coming from wells brought into production after 1972 or from existing wells when production exceeded what was produced in the same month in 1972. The price of new oil was not controlled. Before the embargo the difference in the selling price of "old oil" and "new oil" was less than a dollar a barrel. Within five months, the embargo led to a fourfold increase in the price of imported oil. Unregulated domestic production, that is, "new oil," followed the OPEC price. In September, 1973, new oil was selling for approximately $5.12 a barrel, and by December it was priced at $10.35 a barrel (Congressional Quarterly, Inc., 1981, p. 32).

There was general support in Congress for the oil price regulation implemented by President Nixon in 1973. Price regulation led to a new controversy, however. The limitations on prices were seen as contributing to shortages in some sectors of the economy because the industry channeled its crude oil into the most profitable markets as determined by the regulatory program. Many legislators were unhappy with the way scarce petroleum was being allocated. They wanted the government to assure that priority needs such as health care and agriculture were met (U.S. Congress, 1973, p. 18). There were repeated allegations that the major oil companies were using the shortage to squeeze out independent retailers (U.S. Congress, 1973, p. 9).

In response to these problems with fuel-oil and gasoline allocations, the Congress had initiated efforts to frame new legislation even before the embargo. By November, 1973, a month after the embargo began, Congress had passed emergency legislation, the Emergency Petroleum Allocation Act (EPAA). Initially the act simply required a continuation of the regulatory scheme in place under the existing wage and price controls. Although the EPAA was a response to concerns over allocation, it was to become the primary legal basis for the nation's control of oil prices for the next several years. It also became the basis for the government's entitlements program, which was aimed at equalizing refiners' acquisition costs for crude oil. This was made necessary by the wide range of crude-oil prices resulting from price regulation.

The debate over oil price regulation began to heat up, fueled by two major issues. With the price of new oil going up in parallel with that of imported oil, oil-industry profits were rising rapidly (*Washington Post,* 1973a, p. D13). Profits became the focus of continuing attention by critics of the oil industry (*Washington Post,* 1973b, p. E4). The critics continued to allege that the profits were, in fact, excess profits. The oil industry was said to be getting rich at the expense of an increasingly deprived and distressed nation.

The other issue, articulated by the oil industry, was that only with significant increases in profit would the industry be able to explore for and produce significantly increased amounts of domestic oil. If the real objective was an escape from dependence on insecure, high-cost foreign oil, then, the industry said, the only answer was to find and produce more domestic oil (*Oil and Gas Journal,* 1971a, pp. 64–69; 1972, p. 36; 1973a, p. 51). The nation had extracted all of the easy-to-produce domestic oil, the industry argued, and now had to seek the portion of the resource that was harder to develop. That meant that the cost of producing oil was going to be much higher, and prices simply had to reflect that fact.

Both President Nixon and, later, President Gerald Ford were in sympathy with the position taken by the oil industry. Their general view was that the marketplace should be allowed to set the price (U.S. President, 1974, p. 72; U.S. President, 1976,

p. 289). Nonetheless, both Nixon and Ford were sensitive to the issue of rapidly increasing oil-industry profits, and both proposed a windfall profits tax on the rapidly rising income of the industry (U.S. President, 1974, p. 72; U.S. President, 1975, p. 40). In sum, both Nixon and Ford wanted oil deregulated and the imposition of a windfall-profits tax. The congressional majority clearly was opposed to that formula.

The year 1975 was a year of continuing debate over how to handle oil price regulation. The EPAA was to expire during that year. President Ford could deregulate oil simply by vetoing an extension of the law (Congressional Quarterly, Inc., 1981, p. 154). Public opposition to the deregulation of oil, however, was so widespread that Ford did not want to take full responsibility for such action. It was clearly the president's desire to have a majority of Congress support deregulation.

After extensive sparring over the terms of an extension of the EPAA, President Ford finally signed a new version of it on December 15, 1975. Not only did the new legislation not decontrol oil, it in fact brought the previously unregulated "new oil" under price control.

Beginning with the 1975 extension of the EPAA, an increasingly complex system of price regulation evolved. For example, in 1976 oil from stripper wells—wells producing less than ten barrels a day—was exempted from price controls (Energy Conservation and Production Act, 1976).

Oil price regulation had two objectives. First, regulation should provide incentives for the industry to search for and produce high-cost new oil or oil that was not economically viable without high prices (such as that produced from stripper wells). Second, the industry should not enjoy excess profits. These conflicting goals led to ever-more-complex price regulation. The government's ability to manage this complex system of price regulation slipped with each addition in a pattern of seemingly exponential modifications. With each new package of controls criticism grew. In response to that criticism, new adjustments in regulation were put in place. Almost everyone became unhappy with oil price regulation.

President Carter brought to the White House a strong commitment to conservation. The Carter administration wanted consumers to become sensitive to the scarcity of oil, and the mechanism for achieving that goal was to raise the price of oil. The Carter approach, however, rejected the Nixon-Ford policy orientation. The previous two presidents had called for deregulating the price of oil and levying a windfall-profits tax. Carter proposed instead to levy an excise tax at the well-head equal to the difference between existing controlled prices and the uncontrolled or OPEC price. The revenues from that tax would be returned to the public through the income tax system. Carter also sought the enactment of a tax on gasoline that would increase each year if gasoline consumption increased. These two taxes were part of Carter's National Energy Program, the most comprehensive energy policy package introduced by a president during the 1970s (U.S. President, 1977d, p. 566). A major objective of the Carter administration's proposals was to encourage conservation, but at the same time the administration had only limited faith in the economic value of conservation. Thus it was reluctant to do anything about price control.

Carter's approach to oil pricing, which involved continued controls with new taxes, was adopted by the House of Representatives but was not acted on by the Senate (Congressional Quarterly, Inc., 1981, p. 192). OPEC was continuing to raise the price of oil, and with the overthrow of the shah in Iran came a doubling in price (Gage, 1979, p. A-1). In response, Carter reversed his position and adopted the strategy that had been put forward by both Presidents Nixon and Ford: he opted for the deregulation of domestically produced oil and a windfall-profits tax.

Under the amended EPAA passed in December, 1975, oil price regulation was mandatory through June 15, 1979. At that time price regulation was discretionary with the president until September 30, 1981, at which time control automatically ceased unless new legislation was passed. Using the EPAA's formula, President Carter established a pattern of gradual oil deregulation starting on June 15, 1979, and leading to total deregulation on September 30, 1981 (U.S. President, 1979, p. 609). At the same time that he announced the decision to deregulate oil, the president called for the passage of a windfall-profits tax.

It is a mark of the changing attitudes both in Congress and in the general population that the major concern with Carter's decision for deregulation was whether it should be announced in advance of the passage of a windfall-profits tax (Congressional Quarterly, Inc., 1981, p. 41). As previously noted, the strong opposition to oil deregulation had eroded during the 1970s. First, there was a general sense that only a substantial increase in the price of oil would lead to significant oil conservation. Second, a growing number of observers had become convinced that only a substantial increase in price would stimulate the vigorous search for new domestic oil that was perceived to be necessary. Finally, there was growing disenchantment with the ability of the federal government to manage an ever-more-complex price-regulation system.

Much of the criticism of Carter's decision to announce deregulation before the passage of a windfall-profits tax was based on the belief that Congress would never enact the tax (Congressional Quarterly, Inc., 1981, p. 42). In early 1980 the Congress fashioned a windfall-profits-tax system based on the regulatory categories in existence before deregulation (Pelham, 1980, pp. 668–69). The tax rate was 30 percent on a well brought into production in 1979 or later. Depending upon the character of the production (for example, stripper wells) and the character of the producer (for example, independent versus major), the tax rate for other production ranged between 30 and 70 percent. These taxes were levied on the difference in price between oil sold on the free market and the regulated price in existence before deregulation.

The windfall-profits tax was to remain in existence until $227 billion had been collected. Unless Congress extended the tax, it was to be phased out between 1990 and 1993. The termination date would extend beyond 1990 if it took longer than that to reach the $227 billion figure (Crude Oil Windfall Profit Tax Act of 1980). The decision to deregulate oil was a clear indication that cheap energy was no longer viewed as a practical goal for United States energy policy and that conservation was to play a larger role. On the supply side, these deregulation actions triggered the most rapid increase in drilling for oil and gas in the nation's history (McCaslin, 1981, p. 145).

Depletion Allowance

The other significant tax development with regard to oil during the period from 1973 to 1980 was the elimination of the deple-

198

tion allowance for major oil producers through the Tax Reduction Act of 1975. Independents producing less than 6,000 cubic feet of gas or 1,000 barrels of oil a day retained a 15 percent depletion allowance, a cut of seven percentage points below the depletion allowance that had been written into law in 1969. The oil industry did retain the tax advantage of being able to write off intangible costs related to oil-and-gas production. Although President Carter proposed tightening the foreign tax credit that allows American companies to write off payments to foreign nations against their United States tax bill, Congress rejected that provision (Congressional Quarterly, Inc., 1981, p. 37). Elimination of the depletion allowance was another indication that cheapness was no longer a major goal of American energy policy.

Restrictions on Imports

In April, 1972, before the imposition of the embargo, President Nixon had eliminated the system that required licenses to import oil (U.S. President, 1973b, p. 389) and had replaced it with a tax on imported oil. Although various proposals were made during the 1970s to levy taxes on imported oil with the intent of driving up the price and therefore forcing conservation, none of these were enacted. The dominant concern of the United States during the 1970s was not to exclude the import of oil but, on the contrary, to increase imports to a level adequate to meet domestic demand. In sum, all barriers to importation died with the embargo.

The Trans-Alaska Pipeline System

No action contributed as much to the increase in domestically produced oil as the passage of the Trans-Alaska Pipeline Authorization Act in 1973. In 1968 the Atlantic Richfield Company had discovered oil on the Alaska North Slope. Within the next few years the oil industry had identified roughly 10 billion barrels of reserves in that area. To move North Slope oil to markets in the lower United States, a consortium of oil companies proposed to build a pipeline from Prudhoe Bay to Valdez, on Prince William Sound. From Valdez the oil would be moved down the coast in tankers.

Fearing damage to tundra and wildlife and potential oil spills from tankers, environmental interest groups challenged the industry's proposal for transporting the oil (Congressional Quarterly, Inc., 1981, p. 133). They won a decision in a federal court prohibiting the secretary of the Interior from issuing construction permits until an environmental-impact statement had been prepared. That statement was completed and released in 1972, and the secretary of the Interior indicated his intention to approve construction. The environmental interest groups responded by appealing to a federal court of appeals, which, in February, 1973, prohibited the issuance of the construction permit because the Mineral Leasing Act of 1920 limited rights-of-way across public lands to a width of no more than 50 feet. The oil industry indicated that it needed approximately three times that width (Congressional Quarterly, Inc., 1981, p. 133).

Following the court's decision, the Congress undertook to write legislation that provided for accelerated construction of the Alaska pipeline. The legislation (Trans-Alaska Pipeline Authorization Act) had been fashioned by the time of the embargo, and President Nixon signed it into law in November, 1973.

The key provisions of the act lifted the 50-foot limit on pipeline rights-of-way and barred future challenges to the pipeline under the National Environmental Policy Act. By 1980 the Trans-Alaska Pipeline System (TAPS) was delivering nearly one and one-half million barrels a day to the United States market (Beck, 1981, p. 127). Yet even with the Alaska production, total domestic production in the United States was lower in 1980 than it had been in 1973 (U.S. Department of Commerce, 1981, p. 578). Without the Alaska production, the domestic supply would have dropped precipitously.

In the case of TAPS the Congress and the president decided to overrule environmental concerns and move as rapidly as possible to increased oil production. A review of the struggle over TAPS suggests that there was no very compelling evidence that the pipeline would pose a serious environmental threat. Certainly immediately after the embargo the availability of oil was judged to be vastly more important than what many considered to be rather speculative concerns about environmental damage resulting from the pipeline. Moreover, a great many of the environmental issues were factored into the ultimate design of the TAPS facility.

Federal Lands

Controversy over the appropriate level of energy and mineral development on federally controlled lands has been a continuing issue in American politics. Certainly that conflict concerns much more than oil. It was, however, the embargo and the continuing concern with energy policy that raised the conflict to a high level during the 1970s. One reason for the intensity of the issue can be seen in the view expressed by Robert Nanz, a vice-president of Shell Oil Company, in an address to the Society of Petroleum Engineers in 1980. In the address Nanz said, "It should come as no surprise . . . that 70 percent of the remaining oil and gas to be found in the U.S. is estimated to be on federal lands" (Nanz, 1980, p. 3). Many experts on oil-and-gas resources share that view. Thus, by general consensus a very substantial portion of future domestic oil must come from federally controlled lands. Between 1973 and 1980 major decisions were taken to open and accelerate exploration and development on public lands, specifically the outer continental shelf (OCS), onshore lands in the forty-eight contiguous states, the Naval Petroleum Reserves, and onshore Alaska.

OCS Lands

The OCS includes those lands under federal jurisdiction off the coasts of the United States. In his energy address, delivered in April, 1973, a few months before the embargo, President Nixon called for tripling the amount of acreage leased every year on the OCS (U.S. President, 1973b, p. 389). Two concerns underlay Nixon's focus on OCS lands. First, resource analysts believed that there was great potential for the discovery of large quantities of oil and gas on these lands, and only a very small portion had been leased and thus made available for exploratory drilling. Second, following the Santa Barbara oil spill in 1969, the leasing of OCS lands had been severely disrupted. By 1973 it had not been possible to reestablish a stable, predictable pattern for leasing the lands.

After the Santa Barbara oil spill the Department of the In-

201

terior had significantly modified its regulatory program for OCS oil-and-gas operations with the intent of assuring safer operations and reestablishing public confidence in the safety of those operations (U.S. Department of the Interior, 1980b). Beginning in 1973, in a sustained legislative effort, Congress began working out amendments to the Outer Continental Shelf Lands Act of 1953, the legislative authority for leasing OCS lands, which would provide a new consensus for accelerated development of offshore oil and gas. Enacted in 1978, the Outer Continental Shelf Lands Act Amendments tightened environmental controls, sought to give oil operators other than the major companies increased access to the OCS, and substantially expanded state participation in federal decisions governing oil-and-gas operations.

With the passage of this legislation and the modified regulatory arrangements that had evolved in the Department of the Interior, it was possible to establish an OCS leasing program by 1980 that projected greatly accelerated leasing (U.S. Department of the Interior, 1980a, p. 2). By 1980 a relatively stable environment for OCS operations had been reestablished, and lease sales and development were being carried out on a predictable basis. As part of his new leasing program, the secretary of the Interior offered for the first time leases in various OCS frontier areas, including areas off Alaska, off New England, and in previously unleased deep water. Clearly the decision had been made to expand and accelerate the search for oil and gas on the OCS, but only within a complex set of environmental protection mechanisms (U.S. Department of the Interior, 1981).

Onshore Lands

Few areas of federal policy were affected more significantly by the environmental movement than the government's management of onshore federal lands. These pressures in the end led to the passage in 1976 of the Federal Land Policy and Management Act (FLPMA). The act required that the federal government, specifically the Department of the Interior, carry out a vast program of systematic planning and management for federal lands. The most controversial part of the FLPMA was the directive that the DOI recommend certain areas to be designated as wilderness (energy and mineral development is significantly restricted in or excluded from most designated wilderness areas).

Before the passage of this planning-management legislation there had been a slowdown in oil-and-gas development on federal lands. Although that slowdown continued following the passage of the act while the DOI was carrying out the mandated studies, by the end of 1980 all but 24 million acres of federal lands were released from any special restrictions on their use (Congressional Quarterly, Inc., 1981, p. 104). Although some potentially rich oil-producing areas were contained within the restricted 24 million acres, the decisions under FLPMA opened up vast portions of federal lands and established a more stable, predictable environment.

As was the case with the OCS, the president and Congress clearly had decided that the opportunities for finding and producing oil and gas on onshore public lands had to be significantly enhanced. Again, to many observers the compromises worked out seemed to provide adequate environmental protection while opening up the opportunity for accelerated oil-and-gas development.

Naval Petroleum Reserves

The same wave of change with regard to oil-and-gas development on public lands was reflected in the handling of the Naval Petroleum Reserves (NPR). During the 1920s, Congress had set aside areas in Wyoming, California, and Alaska as Naval Petroleum Reserves (Naval Petroleum Reserves Production Act of 1926). The purpose was to hold these oil-producing or potentially oil-producing areas as a source of oil for the Navy in case of war. The best known of these reserves, Elk Hills, California, was capable of producing over 100,000 barrels a day of oil within a relatively short period upon authorization by the Congress (U.S. Congress, 1976, p. 7).

In the eyes of many observers the nature of modern war—that is, war in which nuclear weapons are employed—made the idea of withholding oil until a war started seem anachronistic. Further, it was widely believed that the nation was in fact faced with a national crisis of major proportions. In 1976, Congress voted to change the strict reserve status of NPRs (Naval Petroleum Reserves Production Act). The reserves in California and Wyoming were left in the hands of the Navy,

but Congress authorized the Navy to begin production immediately. The largest of the reserve areas was in northwestern
Alaska. Its designation was changed from NPR Number Four
to the National Petroleum Reserve in Alaska (NPRA), and its
control was transferred to the DOI. Interior continued an exploratory drilling program begun by the Navy, and by 1980
twenty-eight exploratory wells had been drilled on the NPRA.
With the exception of one small gas field, however, there was
no commercial production (*Oil and Gas Journal*, 1981, p. 51).
In 1980 the Congress authorized the DOE to open the NPRA
to leasing and therefore to private exploration and development (Bureau of National Affairs, 1980a, p. 11).

Onshore Alaska

Congressional authorization to lease on NPRA followed immediately on the heels of the passage of the Alaska Lands Act
in 1980. This legislation was perhaps the most controversial
federal-lands bill undertaken in the 1970s. It set aside approximately 104 million acres of federal lands in Alaska as parks,
refuges, and other types of conservation areas. In many of these
areas energy development was restricted; however, most of the
areas thought to have the greatest potential for oil-and-gas production lay outside these areas.

Finally, and of perhaps greatest importance to energy, there
had been almost no oil-and-gas leasing on federal lands onshore
in Alaska for the decade preceding the passage of this act. With
its enactment vast areas of Alaska were available to the oil
industry for leasing, exploration, and development.

Although the movement occurred with many fits and starts
during the 1970s and throughout 1980, the clear wave of change
was in the direction of making more public lands available for
more rapid leasing, exploration, and production of oil in every
area. On the other hand, the legislation and policy actions with
regard to federal lands established a pattern of environmental
control that was the source of continuing unhappiness among
many people in the oil industry, as well as others concerned
with the most rapid possible development of domestic United
States resources. An overview of what occurred, however, makes
it clear that the dominant issue with regard to energy on public
lands was that of security. Threatened by another oil shortfall,
the nation was moving in a way that it had not moved in

decades—to make public lands available for energy development.

Short-Term Responses

The preceding sections have described the actions taken by the president and Congress to address the long-term problem of inadequate domestic supplies of oil. The embargo and the "second shock" disruption resulting from the revolution in Iran caused serious, though short-term, scarcity problems. By 1980 strategies were in place for dealing with short-term disruptions of petroleum: a system of rationing (Emergency Energy Conservation Act, 1979) and the Strategic Petroleum Reserve (Energy Policy and Conservation Act, 1975).

One of the elements of the EPAA of 1973 was to require the president to set up a system for allocating petroleum to assure that petroleum was distributed equitably. The embargo, which lasted approximately six months, had ended before anything more than a skeleton allocation system had been put in place. One lesson was learned from this experience: it was extremely difficult to establish in a short time a federal allocation system that effectively managed the complex petroleum system.

The Iranian oil disruption led to the passage by Congress of a gasoline-rationing program (Emergency Energy Conservation Act of 1979). Under this legislation the president could call for rationing only if a shortage of 20 percent had existed for thirty days. Under those circumstances the president could act only if neither house of Congress vetoed the decision to implement rationing. Clearly any rationing program was going to be even more complex to manage than was an allocation program. Little enthusiasm existed for rationing, but by 1980 the nation had a program that was something like a gun behind the door.

The more significant response to the short-term problem was the decision in 1975 to establish a Strategic Petroleum Reserve which would ultimately store one billion barrels of oil (Energy Policy and Conservation Act, 1975). Although there was widespread support for a reserve that could carry the nation over a short-term denial of oil imports, by early 1979 the nation had only approximately 70 million barrels of oil in the SPR (Congressional Quarterly, Inc., 1981, p. 100). Further, the oil was not available because no pumps were in place to extract it. The problems of implementing the SPR turned out to be multiple, ranging from unexpected engineering difficulties associated with storage in the salt domes in Louisiana and Texas to opposition by Saudi Arabia to increasing estimates of the overall costs (Congressional Quarterly, Inc., 1981). The engineering problems arose from lack of experience with a reserve of this size, to be established on such an accelerated schedule. The Saudis objected because they were trying to keep down world prices and saw the filling of the reserves as increasing pressure for price rises. The financial problem resulted from the fact that after the Iranian-triggered price increase one billion barrels of oil would cost $35 billion.

In 1980 the Congress passed legislation mandating that the president put into the SPR at least 100,000 barrels of oil a day (Energy Security Act, 1980). The Congress built into that legislation an enforcement mechanism. It specified that if the 100,000-barrel-a-day level was not achieved within six months oil was to be diverted from Elk Hills NPR to the SPR.

Research and Development

Although the nation as a whole responded to the energy crisis by accelerating the flow of federal money into various research-and-development activities, very little federal funding went into oil-related R&D. On the one hand, the petroleum industry was suspicious of federal R&D funds. There was concern in the industry that federal involvement might ultimately lead to some kind of federal oil-and-gas organization. On the other hand, many members of Congress believed that the petroleum industry was receiving sufficient income through the higher price of oil to cover any needed R&D activities.

Perhaps the most significant federal activity in this area was the funding of a major test program aimed at improving the industry's ability to carry out tertiary oil recovery. One of the possibilities for increased domestic production was thought to rest with the oil that remains in reservoirs after conventional production has come to an end. It is generally estimated that only one-third of the oil in the ground is produced. The goal of the Tertiary Recovery Test Program was to find techniques to increase that recovery and therefore increase domestic production. By 1980 the test program was coming to an end, and clearly no enhanced-recovery panacea had been found.

Summary

By 1980 the president and the Congress had initiated the deregulation of oil. The deregulation pattern had resulted in the most substantial efforts at oil conservation that had occurred following the oil embargo. Additionally, the deregulation had stimulated an accelerating rate of drilling. By the end of 1980 drilling was occurring at the highest rate in history. To recover oil on federal lands, the federal government had moved across the board to make those lands more rapidly available for oil exploration. The nation had in place a rationing system and late in 1980 had begun filling the Strategic Petroleum Reserve on an accelerated basis as a result of congressional mandate.

NATURAL GAS

In late 1978 the Congress and the president reversed a trend toward ever-more-detailed federal administrative regulation of the price of natural gas. The passage of the Natural Gas Policy Act of 1978 reversed roughly forty years of policy evolution emanating from the courts and the Federal Power Commission (FPC). The move toward gradual deregulation of natural gas was an event of great symbolic as well as practical importance. Gas deregulation signaled the reversal of the nation's assumption of a future characterized by cheap energy.

The action on price deregulation for natural gas, however, was not as clean a break with the past as was President Carter's decision the following year to deregulate oil. Rather, the gas-deregulation statute was a very complex, highly detailed piece of legislation that directed only partial decontrol of prices.

First, the Natural Gas Policy Act brought intrastate gas under federal price controls. Second, it designated that all gas that had come into production in 1977 or before would remain regulated. This "old gas" would be regulated at a price higher than the previous price and would be allowed to increase with inflation. Third, it established a deregulation schedule for "new gas" that came into production after 1977. Deregulation of "new gas" would occur by 1985 and would involve a series of steps whereby the price of gas would rise at a rate roughly 4 percent higher than the rate of inflation. This new gas would be totally deregulated in 1985 unless the president and Congress took action to continue the regulation. Fourth, the act specified that certain categories of "high-cost" gas would be deregulated about one year after the enactment of the legislation. "High-cost" gas, for example, included gas produced from wells deeper than 15,000 feet. Perhaps the complexity of the legislation can be most easily communicated by noting that there were more than twenty price categories for the gas that was to continue under regulation until 1985, the price paid for gas in each category to be adjusted on a monthly basis in response to inflation. The Natural Gas Policy Act was thus at once a fundamental break with forty years of government policy and an extension of a very complex regulatory system.

In his April, 1973, energy policy address, President Nixon called for the deregulation of gas from new wells, gas newly dedicated to the interstate market, and gas newly available because of expired contracts (U.S. President, 1973b, p. 389). Nixon noted that the result of the artificially low regulated price of interstate gas had been that 50 percent of the nation's natural gas was consumed by industrial users and utilities. He suggested that without such regulation many of the industries and utilities would have been using coal or oil. Further, Nixon noted the difference between intrastate and interstate prices. In sum, the president said, price regulation had created excessive demand for gas and, in addition, had eliminated any incentive for companies to find new gas. These issues, plus the concern that deregulation would generate excess profits for producers and unacceptable financial burdens for consumers, were the essence of the debate over gas regulation that continued until 1978.

In the early 1970s the growing shortage of gas in the interstate system brought unremitting pressure for some kind of modification in gas price regulation (Oil and Gas Journal, 1970b, p. 52). The FPC had responded to the impending gas shortage by raising regulated prices (Oil and Gas Journal, 1970a, p. 40). As was the case with oil regulation, each step in the FPC's process was aimed at protecting against a windfall profit on gas that had been brought into production before energy prices began to rise while also providing incentives for the discovery and delivery of "new gas" to the interstate market.

Again as was the case with oil, however, each effort to balance concern about excess profits with the need to provide incentives resulted in a more complex regulatory system. The inability to find a regulatory formula that worked caused more and more people to question the efficacy of gas price regulation.

The severely cold winter of 1976–77 was the turning point in widespread recognition that some fundamental change had to be made in the existing price regulation system for natural gas. The northeastern part of the United States found itself with inadequate supplies of gas (though this may have been the result as much of deficiencies in the transmission system as of any real shortfalls in supply). The shortage led to the closing of schools, shutdowns of a number of industries, and under-heated homes. The FPC responded under emergency provisions and allowed interstate pipelines to buy gas for prices far above regulated levels. Under this authority, however, these sales could be permitted for no more than sixty days (Congressional Quarterly, Inc., 1981, p. 184).

When President Carter was inaugurated in January 1977, the gas situation in the Northeast was severe. Within two weeks of his inauguration Congress had passed and he had signed the Emergency Natural Gas Act of 1977. Under that legislation the president was given temporary authority to transfer interstate gas supplies to areas in need and to approve sales of gas to interstate buyers at unregulated prices. The law was aimed at dealing with the crisis of that winter on a short-term basis.

During the previous year growing recognition of an impending gas shortage had led Congress to pass legislation setting deadlines for federal decisions on how best to transport gas from the North Slope of Alaska to the lower forty-eight states (Alaska Natural Gas Transportation Act, 1976). As was true with oil, there were huge proven gas reserves in the North Slope, but there was no way to make those reserves available to the United States market (U.S. Congress, 1977a, pp. 2–3).

Three alternative gas-pipeline proposals were pending before the FPC at the time Congress acted in 1976. The key element in the action by Congress was to require that the FPC recommend by May 1, 1977, whether to proceed with a gas-transmission system and, if the recommendation was positive, determine the preferred system. The president was to send his recommendation to Congress by September 1, 1977.

Recalling the gas shortages of the previous winter, the Congress acted rapidly in passing a resolution of approval of President Carter's choice of an Alaskan gas pipeline (Alaska Natural Gas Transportation System Approval, 1977). Although no construction had begun on the pipeline by the end of 1980, the selection of a pipeline system, estimated to cost $15 billion, was yet another sign of the growing recognition that natural-gas prices could no longer be maintained at the low levels of the past.

As befitted a decision that symbolically brought to an end a two-hundred-year history of cheap energy, the struggle over the Natural Gas Policy Act was long and tortuous. James Schlesinger, Carter's secretary of energy, is reported to have said about the congressional struggle: "I understand now what Hell is. Hell is endless and eternal sessions of the Natural Gas Conference" (Congressional Quarterly, Inc., 1981, p. 51). By the end of 1980 the Natural Gas Policy Act had by general agreement accomplished what its advocates had said that it would accomplish. The gas shortage disappeared. There was accelerated drilling for natural gas, and large quantities were being found. Even the complexity of the act seemed to be paying off. The total deregulation of high-priced gas had led to accelerated efforts to drill for gas below 15,000 feet. In the Anadarko Basin, in western Oklahoma, there was early evidence of huge quantities of natural gas. The fact that the NGPA allowed the gas pipelines to balance the price of this unregulated gas with the lower-priced regulated gas meant that some of the deep gas sold for approximately $9.00 per thousand cubic feet while the average price for regulated gas was about $2.00 to $2.50 per thousand cubic feet.

Summary

New exploratory, discovery, and drilling activities prompted by the partial, graduated deregulation of natural gas led some observers to believe that it could serve as a transition fuel; that is, gas could carry the nation from the energy mix of the years before the embargo to a set of more esoteric, new-era energy sources that would come on line in the post-2000 years. It should be noted, however, that powerful forces in the United

States were still pushing for total gas deregulation, and Ronald Reagan in his 1980 election campaign indicated his commitment to total deregulation of natural gas.

COAL

Price regulation was not an issue in the search for a national policy for coal. The policy debate concerning coal was not, therefore, as highly charged as that focused on oil and gas. It was infinitely more complex, however. Because of the nation's great abundance of coal, every president from 1973 onward focused attention on expanding the place of coal in the nation's energy system, and there were continuing expressions of frustration about the nation's slow-paced shift to increased utilization of coal.

In fact, however, between 1973 and the end of 1980 substantial progress had been made in formulating policy for coal. Coal production in the United States rose by nearly 40 percent in that seven-year period. In 1973 production was 599 million tons, whereas in 1980 production totaled 835 million tons (U.S. Department of Energy, 1982, p. 123). Coal's overall role in the nation's energy system rose from 17 percent in 1973 to 21 percent in 1980 (Bureau of National Affairs, 1982, p. 819).

Coal policy initiatives in the seven-year period fell into five categories: (1) environmental protection, (2) utility conversion, (3) mining on federal lands, (4) research and development, and (5) synthetic fuels.

Environmental Protection

One of the first responses to the rapidly rising concern with environmental quality in the 1970s was the passage of the Clean Air Act of 1970. By that legislation the government was required to establish both primary and secondary air-quality standards. Primary standards were to protect human health, and secondary standards were to protect the general welfare. In his energy message to the Congress of April 18, 1973, President Nixon called for a slowdown in the application of the secondary standards (U.S. President, 1973b, p. 389). Nixon said that any effort to establish secondary air standards by 1975 would

211

prevent the use of up to 155 million tons of coal a year (*Congressional Record*, 1973, pp. 576-94).

The debate over the stringency of air-quality standards continued through the 1970s. Opponents argued that strict air-quality standards denied the nation the opportunity to use larger quantities of coal and that, in turn, resulted in the importation of larger quantities of oil. Proponents argued that strict standards were central to achieving acceptable levels of air quality.

In 1977 the Congress amended the Clean Air Act and added to the stringency of the Federal Air Quality Regulation Program. By 1980 the Environmental Protection Agency was trying either to enforce or to formulate air-quality regulations that established national standards for air and special standards for pristine areas (known as Prevention of Significant Deterioration Areas) and to implement standards for air emissions from large plants (known as Best Available Control Technology Standards). Although there was continuing controversy about the precise character of the standards and how they should be enforced and the various administrations modified the administrative and technical approaches, the overall movement of air-quality regulation through 1980 was toward stricter control of air pollution, and coal was the predominant fuel affected by those regulations.

Most air-quality regulation of the 1970s focused on three pollutants: sulfur dioxide, nitrogen oxides, and particulates. By 1980, however, concern was growing about acid rain, acid dust, carbon dioxide, and other pollutants (Congressional Quarterly, Inc., 1981, p. 110). There was much controversy within the scientific-technical community about appropriate standards, appropriate control technologies, and the most cost-effective approach to regulating all these pollutants. The scientific debate fed the controversy over air-quality regulation. The nation's approach to coal was clearly that it should be used but only when appropriate measures had been taken to protect the environment.

Much the same evolution can be seen in the choices made about the control of strip mining. Concern over strip mining had been building for many years. In 1972 the House of Representatives had approved a bill setting federal strip-mining standards (*Congressional Record,* 1972, p. 35031). In 1974 both the House and the Senate approved a strip-mining bill, but that legislation was pocket-vetoed by President Ford (Congressional Quarterly, Inc., 1975, p. 579). Ford vetoed another strip-mining bill in 1975 (U.S. Congress, 1975, *House Doc.* 94–160, pp. 3–4). With the arrival in the White House of President Carter, the Surface Mining Control and Reclamation Act of 1977 was passed and signed by the president.

The new legislation established national standard and a new Office of Surface Mining and Enforcement (OSME) in the Department of the Interior to implement the legislation. The OSME prepared a complex set of regulations requiring mine operators to meet very detailed standards. The industry challenged these standards in court, and mine reclamation remained controversial through 1980 (Andrus, Secretary of the Interior v. Virginia Surface Mining and Reclamation Assn., et al.).

With the overthrow of the shah of Iran came new pressures to accelerate the development of domestic energy, particularly coal. In this context President Carter, who was generally perceived as an environmentalist, proposed to the Congress the establishment of a federal Energy Mobilization Board (EMB). This board was to have the authority to place key energy projects on a "fast track." The rationale behind the EMB was the perceived need to overcome the complex bureaucratic and regulatory barriers to the construction and operation of new energy facilities. To the surprise of many observers, however, the Congress killed this legislation. The EMB was opposed by environmentalists who feared that it would negate many of the environmental regulations of the 1970s. The major allies of the environmentalists were conservatives who were fearful that the board would both infringe on state's rights and create another burdensome federal bureaucracy (Congressional Quarterly, Inc., 1981, p. 106).

Thus in the period between 1973 and the end of 1980 the thrust of federal policy with regard to coal was to find a way to expand the use of the resource while at the same time protecting against the adverse environmental consequences of its use. Time after time, however, when the Congress was faced with a choice, it came down on the side of environmental protection, as reflected in the choices made with regard to air quality, surface mining, and the rejection of the Energy Mobilization Board.

Utility Conversion

When President Nixon called for utilities to switch from oil to coal where that was technically possible in his message to Congress of April, 1973 (U.S. President, 1973b, p. 389), he reflected a growing perception that it was not in the national interest to use the premium fuels, oil and gas, to generate electricity. The loss in that conversion process of roughly two-thirds of the input energy of oil and gas, plus the fact that large utility and industrial boilers offered the greatest opportunity for coal cleanup, made utility conversion to coal particularly attractive. With the passage of the Energy Supply and Environmental Coordination Act in 1974 the Congress made its first move to require utility conversion to coal. The act, however, had so many exemptions that it was only minimally effective in forcing conversion. Many utilities resisted federal pressures to convert to coal because of its complex handling requirements, the environmental problems associated with coal use, and unstable management-labor relations in the industry. The 110-day strike by the United Mine Workers in 1978 only reinforced the latter concern.

A major element of the National Energy Program that President Carter submitted on April 20, 1977, was the proposed use of a complex system of taxes and regulations to force the utility industry to move to coal. An extensively modified version of the proposals in the National Energy Program was ultimately enacted into law by the Congress in the Powerplant and Industrial Fuel Use Act of 1978. That legislation directed utilities using gas and oil in coal-capable boilers to switch to coal. Although many exemptions were allowed, the act was very specific in banning the use of natural gas by utilities after 1990. In general the utilities appeared to be moving to coal as the energy source for their new plants after 1978, but they continued to push for exemptions from the 1978 law requiring conversion of existing oil- and gas-burning plants.

In 1980, President Carter sought to achieve the same conversion objective through a package of federal financial inducements rather than sanctions (U.S. President, 1977a, p. 448). Under that proposal the federal government would have paid utilities that were in financial difficulties to convert to coal. The Congress, however, rejected the Carter proposals.

By 1980 the efforts to require or induce utilities to move from oil and gas to coal had reached the point where large new utility plants were based totally on fuels other than oil and gas. The declining confidence in nuclear power as an energy source meant that the overwhelming majority of the planned new plants would be built to use coal as the primary energy source. This federal push toward conversion to coal was very strongly reinforced by the rapidly rising cost of oil and gas following the revolution in Iran, the passage of the Natural Gas Policy Act, and the decision of the Carter administration to deregulate domestically produced oil.

Mining on Federal Lands

By 1980 the Department of the Interior had established a leasing program for mining coal on federal lands and had published a five-year leasing schedule. The first general coal-lease sale in ten years was held in the early part of 1980.

Public lands are estimated to contain approximately 35 percent of the nation's coal reserves (Bureau of National Affairs, 1980c, p. 1107). A large quantity of those reserves can be recovered by low-cost surface mining, and much of the coal is "clean," with a low sulfur content. The focus on coal following the embargo made increased access to federally owned coal a major policy concern.

Although there was a general commitment to increase the availability of federal coal, conflicts arose over the appropriate leasing arrangements. In 1976 the Congress had passed the Federal Coal Leasing Amendments Act and had then overridden President Ford's veto of the act. The legislation required that all coal leases must be issued on a competitive-bid basis, increased the minimum royalty rate, and required lease owners to meet diligence requirements, that is, to mine all of the recoverable reserves on the lease within a forty-year period.

With the arrival of the Carter administration one of the early actions taken by the DOI was to formulate a detailed regulatory system to govern federal coal leasing. Constructing

215

the regulatory program was a complex task because the leasing program had to be consistent with the Federal Land Policy and Management Act, the Surface Mining and Reclamation Act, and many other environmental requirements. Secretary of the Interior Cecil D. Andrus established a deadline of 1979 for the department to formulate its leasing program, and that deadline was met.

Although unhappiness was widespread in the coal industry over the DOI's regulatory program, the leasing program did break the stalemate that had persisted for a decade. The pattern paralleled that with regard to oil and gas on federal lands. Clearly the decision was to open more federal lands for development.

Research and Development

Expenditures for federal research and development in coal grew rapidly in the period between 1973 and 1980. Before 1973 the limited coal R&D activities under way in the DOI and in the Environmental Protection Agency were aimed at understanding the environmental and health impacts of coal burning and developing coal-cleanup technologies.

By 1980 the support for R&D in coal was broad-based and approached $1 billion a year. Research included programs in the DOI, primarily in the U.S. Geological Survey, to identify and characterize the nation's coal resources through a wide-ranging program in the Department of Energy, and environmentally oriented activities in the EPA. Coal R&D expenditures by government were second only to those for nuclear power. By 1980 federal funds were supporting R&D in every generic approach that had been proposed for cleaning coal, burning it more efficiently, and converting it into other fuel forms.

By 1980, R&D expenditures by the EPA had played a major role in the development of flue-gas desulfurization technology, and cleanup technology was being required as a component of large new coal-fired electric power plants. The rapid growth in federal R&D in coal was motivated by a desire to find environmentally acceptable ways of using the nation's abundant coal resources, and also converting it into higher-quality fuels, specifically liquids and gases.

216

Synthetic Fuels

The largest single federal expenditure commitment on coal was aimed at advancing the technology for creating synthetic fuels (synfuels) from coal. A number of relatively small-scale pilot and demonstration synthetic-fuel plants had been built during the 1970s as cooperative ventures between the federal government and private industry (U.S. Congressional Research Service, 1980). By 1979 there was broad support in the Congress for a major national effort to commercialize synthetic fuels. One of the justifications offered by the Carter administration for the windfall-profits tax was that a portion of the revenues would be applied to commercialization of synfuels.

President Carter first proposed authorizing $88 billion to push private industry into the synthetic-fuels business. The Congress modified his proposal and in 1980 authorized an initial $20 billion for the federal Synthetic Fuels Corporation (SFC; Energy Security Act, 1980). The legislation specified that Congress could appropriate an additional $68 billion—the remaining portion of the proposed $88 billion—following expedited procedures.

The synfuels legislation established a goal of 500,000 barrels of oil equivalent a day by 1987, that goal to increase to 2 million barrels a day by 1992. The urgency that Congress placed on synfuel development can be seen in the fact that the DOE was authorized to handle synfuels until the new corporation was in operation. The SFC was to be governed by a board of directors, each member's term to be seven years to insulate the corporation from political pressures and give it the ability to operate with something approaching to the freedom of a private-sector corporation.

The SFC was given the authority to employ a variety of instruments to move the nation toward commercial synfuels. Those instruments included loan guarantees, price guarantees, and purchase agreements. The corporation could, as a last resort, contract with industry to build and operate government-owned plants.

From the time of the oil embargo there had been continuing discussions about a national program, of proportions equal to the Apollo Program, aimed at addressing the nation's energy needs. The SFC represented, at least in broad outline, a com-

217

mitment of that kind. Although the synfuels initiative covered more than coal, it clearly offered for the first time the very real possibility of commercializing synthetic fuels from coal.

Summary

By 1980 the nation had evolved a public policy for coal that had six elements. First, there was a continuing commitment to expand the role of coal in the nation's energy system. Coal production grew by approximately 40 percent during the seven years following the embargo. Second, although coal was consistently pushed as a substitute for other energy sources, there was an ongoing commitment to protect the environment. Air-quality and reclamation standards were stricter at the end of this seven-year period than they were at the beginning, and Congress had refused to establish an Energy Mobilization Board (EMB), in part because of concern that it might dilute environmental protection. Third, Congress had enacted legislation that mandated utility conversion from oil and gas to other fuel sources. Coal appeared likely to be the predominant source of energy for new electric plants in the years following 1980. The primary push behind conversion to coal appeared to be the rapidly rising price of oil and gas. Fourth, the federal government had established a program for leasing coal on federal lands and had held the first general lease sale. Fifth, a nearly $1 billion R&D program in coal was in operation. Finally, the Congress had established and generously funded the Synthetic Fuels Corporation.

CHAPTER 9

The Path to Stability: Nuclear Energy, Hydroelectric Power, Oil Shale, Tar Sands, Geothermal Energy, Organic Wastes, Solar Energy, Electricity, Energy Conservation

Policy for the three fossil fuels covered in the preceding chapter dominated much of the presidential-congressional attention between 1973 and 1980. It was, however, the decisions with regard to the other nine energy sources that set the overall boundaries necessary for replacing the fuel policy systems with an energy policy system. In the following the presidential-congressional actions taken for those other nine sources are covered.

NUCLEAR ENERGY

Unlike the policies for the three conventional fossil fuels, which in the period between 1973 and the end of 1980 made significant strides toward a national consensus, policy for nuclear power was in near-total disarray by the end of 1980. A *Congressional Quarterly* publication characterized the situation as follows:

Congress had not given clear directions to the Executive Branch in years about issues such as disposal of radioactive waste, the extent to which costly safety features should be required, changes in licensing procedures, standards for radiation exposure, allocation of the costs of accidents, and decommissioning of plants. In addition, proponents of nuclear power still sought a hardy endorsement from Congress, while opponents looked for a vote halting expansion until controversies such as waste disposal had been resolved. However, the majority in Congress apparently fell in neither camp as 1981 began. These moderates didn't want to abandon nuclear power, but they wanted operation of reactors and disposal of waste to be as safe as possible. Exactly what that meant, they weren't sure. (Congressional Quarterly, Inc., 1981, p. 80)

As discussed in chapter 6, nuclear power involves a nuclear fuel cycle. It is nearly impossible to discuss one component of that cycle independent of the whole cycle. So it is with any effort to understand the policy process with regard to nuclear power.

Projections about the future role of nuclear power declined precipitously between 1973 and the end of 1980. In 1972 the government estimated that the nation would have the equivalent of twelve hundred 1,000-megawatt nuclear power plants by the year 2000. By the end of 1978 those estimates clearly were beyond the limits of political and technical feasibility (Bupp, 1979, p. 126).

Following the cutoff of foreign oil, nuclear energy supporters had high hopes that the oil shortage would be a major impetus for the development of nuclear power. That hope lasted for an exceptionally short time. The key factor in the precipitous decline of federal projections for nuclear power was the change in attitude of the electric utility industry. During the late 1960s and early 1970s the utility industry placed an ever-growing number of orders for nuclear power plants. By the late 1970s the utilities not only were not placing new orders but were canceling earlier decisions and halting construction of plants that had already been started (Bureau of National Affairs, 1980c, p. 17). By the late 1970s the economics of the utility industry, once so predictable, had become an unpredictable morass. First, growth in demand for electricity was down precipitously and unexpectedly. Interest rates had made the cost of borrowing capital a major burden for the utilities, and cost overruns on nuclear plants appeared to escalate with each passing year. All these uncertainties were exacerbated by the inability to resolve three pressing nuclear energy policy issues: (1) nuclear safety, (2) nuclear wastes and spent fuel, and (3) the nuclear future.

Nuclear Safety

Even before the oil embargo, concern about the safety of nuclear

220

power plants had reached the point where the Atomic Energy Commission had decided to commission a major study. The study, directed by Norman C. Rasmussen, of the Massachusetts Institute of Technology, was completed late in 1975 (U.S. Nuclear Regulatory Commission, 1975). The Rasmussen report was a large, complex effort, later described by another safety study group, the Risk Assessment Review Group, as "inscrutable to peer review" (Congressional Quarterly, Inc., 1981, p. 84). The report concluded that, while power-plant accidents were possible, they were exceptionally improbable.

Although the Rasmussen report prompted a small industry in critiques, the Congress at first found it sufficiently reassuring to extend the Price-Anderson Act for ten years in December, 1975 (Atomic Energy Act, 1974).

It will be recalled that the Price-Anderson Act was passed in 1957 to assist the nuclear industry, which was unable to obtain unlimited liability coverage from private insurance carriers. The legislation limited the industry's liability to $560 million for a single accident and committed the government to insure the difference between the amount of insurance available from private insurers and the $560 million limit.

Critics of nuclear safety continued to argue that if nuclear power was safe it should be possible for the industry to acquire private insurance coverage. The extension of Price-Anderson was, however, overwhelmingly supported by the Congress and signed into law by President Ford.

From 1975 onward nuclear safety was an ever-present and growing issue. On March 21, 1978, the Carter administration sought to speed up the licensing of conventional nuclear reactors by proposing legislation that would shorten the time between the decision to build a nuclear power plant and its licensing to six and one-half years from the ten to twelve years that was by that time the norm (Congressional Quarterly, Inc., 1981, p. 203).

The effort by the Carter administration to shorten the time needed to bring conventional reactors on line paralleled initiatives that had been made previously by Presidents Nixon and Ford (U.S. President, 1973a, p. 1319; U.S. President, 1976, p. 289; U.S. President, 1977c, p. 571). The Carter administration's support of accelerated licensing procedures reflected a view that conventional nuclear power had to be retained as a viable option in a nation short of energy. The proposal for accelerated licensing, however, was not acted on by the Congress. Concern over safety continued to be an impediment to the expansion of nuclear power.

In March, 1979, the nuclear accident at Three Mile Island raised even more questions about nuclear safety. President Carter established a presidential commission to study the accident and make recommendations. On October 30, 1979, the Commission, chaired by John Kemeny, president of Dartmouth College, issued its report (U.S. Congress, 1979). The report was the most critical government report ever issued on nuclear power. In response President Carter proposed a number of changes in the structure of the Nuclear Regulatory Commission, the agency regulating nuclear safety. The changes that came out of the Kemeny report did nothing to resolve the nuclear stalemate, however. By the end of 1980 there was greater uncertainty over the safety of nuclear power than at any other time in history. This uncertainty contributed to an accelerating rate of nuclear plant cancellations by the utility industry.

Nuclear Wastes and Spent Fuel

The concern with nuclear safety was accompanied by a growing concern over the handling of nuclear wastes and spent nuclear fuel. By the end of 1980 concerns were being expressed that some nuclear reactors would have to be shut down before the end of their useful lives because of the lack of storage and disposal facilities for spent fuel and nuclear wastes (Bureau of National Affairs, 1980e, p. 17).

In the early days of the nuclear power industry it was the expectation that spent fuel and used fuel rods from nuclear reactors would be disposed of in the following steps. Immediately after their removal from the reactors, the used fuel rods would be placed in water tanks at the generating plant and stored there for approximately one year. During that period the radioactivity would degrade. The next step would be to transfer the used fuel rods to a reprocessing facility. There the spent fuel would be separated into fission products and transuranics. The transuranics would be removed from the reprocessing plant and refabricated into new fuel. The fission products would be moved to a permanent storage site for high-level wastes. The permanent storage site would likely be deep under the surface of the earth.

Between 1966 and 1972 a commercial reprocessing plant operated in western New York State. The plant closed in 1972, when operating problems and changes in federal regulations made it unprofitable (Congressional Quarterly, Inc., 1981, p. 236). When the facility was shut down, the consequences were not initially seen as serious. Plans were in the works for construction of two new reprocessing facilities, one at Barnwell, South Carolina, and the other at Morris, Illinois. Hopes that those plants would come into operation ended in 1977, when President Carter banned reprocessing out of fear that the plutonium produced would result in a proliferation of nuclear weapons (U.S. President, 1977a, p. 448). In the absence of reprocessing capabilities and permanent storage, it was only a matter of time until the storage of spent fuel at nuclear reactors would fill the storage pools.

The search for an answer to the high-level waste storage problem produced two options. The Carter administration proposed the development of a permanent geologic repository for high-level wastes. The other option was storage in above-ground vaults for up to one hundred years (Congressional Quarterly, Inc., 1981, p. 90). Those advocating above-ground vaults claimed that stored materials could be monitored and retrieved whenever necessary either for reprocessing or for permanent storage when a site was developed at some point in the future. The inability to reach a compromise on this issue led President Carter to establish in 1980 the State Planning Council under the chairmanship of Governor Richard W. Riley of South Carolina (Bureau of National Affairs, 1980b, p. 9). The council was to develop a strategy for handling nuclear-waste storage that would be acceptable both to the states and to the federal government. By the end of 1980 no progress had been made.

On another tack the nuclear industry was advocating that the federal government take responsibility for the accumulating body of spent fuel. The proposal was that the federal government develop away-from-reactor (AFR) facilities in which to store the spent fuel that utilities did not have room for at their nuclear generating plant sites (Bureau of National Affairs, 1980d, p. 8). Again, no resolution was achieved on the AFR facilities.

Perhaps the clearest indication of the state of national policy with regard to nuclear wastes is shown by the fact that the Congress was able to enact only two pieces of legislation in the period between 1973 and the end of 1980. Both laws were passed in 1980. One made disposal of commercial low-level wastes the states' responsibility (Low Level Radioactive Waste Policy Act, 1980). Low-level wastes consist of contaminated protective clothing, rags, vials, and so on, as well as radioactive materials used in medicine. When the act was passed, only three states had facilities for disposing of commercial low-level wastes, and all three had either closed down or severely restricted the disposal of low-level wastes from other states.

The second piece of legislation dealing with nuclear wastes permitted the federal government, jointly with the state of New York, to remove high-level liquid radioactive waste stored at the site of the closed nuclear reprocessing plant in western New York (West Valley Demonstration Project Act, 1980). Wastes were to be solidified and moved to a federal repository. The driving force behind this legislation was fear that the two tanks that were holding the waste would develop leaks.

By the end of 1980, then, the nation's inability to find a means of storing nuclear wastes and managing spent fuels not only inhibited the construction of new nuclear plants but made it likely that existing plants would have to be shut down when on-site storage capacities were reached.

The Nuclear Future

The same deadlock that existed with regard to nuclear safety and nuclear wastes impeded the development of the second generation of nuclear power, the fast breeder reactor. In 1970 the Congress had authorized construction of a demonstration breeder reactor on the Clinch River, in Tennessee (Appropriations, Atomic Energy Commission, 1970). It was expected to demonstrate to the electric utility industry the advantages of the breeder system.

By the mid-1970s a swelling chorus of criticism was aimed at the Clinch River project. The critics argued that rising costs

224

plus safety and environmental implications demanded cancellation of the project. When the project was authorized, it had been expected to cost $700 million. By 1980 almost $1 billion had been spent on design and the purchase of equipment, and no work had been done on the site. It was estimated that the reactor would cost almost $3 billion before it was completed.

Opponents of the Clinch River Project gained a powerful ally with the election of Jimmy Carter. During 1977, President Carter indicated his opposition to the breeder reactor, declared that the United States would not use plutonium as a nuclear fuel, and proposed a budget for the Clinch River plant of only $33 million (Congressional Quarterly, Inc., 1981, p. 185), the money to be used to shut down the project.

The primary reason for Carter's opposition was again concern that the plutonium produced by breeder reactors might be illegally diverted into the production of nuclear weapons. Further, Carter argued, experience with the design of the breeder indicated that the concept was technically obsolete and was not economically viable (U.S. President, 1977e, p. 1726).

Congress rejected Carter's plan to shut down the project, authorizing funding at the $80 million level (Supplemental Appropriations Act, 1978). Carter responded by vetoing the bill. Later in 1978 Carter signed a supplemental funding act containing money for the project but indicated that he would use the money to shut down the project. The comptroller general of the United States informed the president that such use of the money would be illegal, and Carter dropped his plan to shut down the breeder, but he did not change his opposition to it (U.S. Congress, 1977b, p. 1). The effect of the conflict between the president and Congress over the future of the Clinch River reactor was that no energy appropriation measure was passed during either 1979 or 1980. Design work on the project and purchase of hardware did not come to a halt because money was provided under continuing resolutions. Not a shovelful of dirt, however, was turned at the site during this period because the Nuclear Regulatory Commission refused to license the construction of the plant. The hardware purchased remained in storage.

The future of nuclear power did receive one boost during this period. In 1980 the Congress, with the support of the administration, passed legislation authorizing a major acceleration in the development of third-generation nuclear technology, that is, nuclear fusion (Magnetic Fusion Energy Engineering Act, 1980). The intent of the legislation was to accelerate work on fusion reactors by fifteen to twenty years. But while the legislation authorized the expenditure of as much as $20 billion on fusion-reactor work by the year 2000, no funds were actually appropriated.

Summary

By the end of 1980 the issues surrounding nuclear safety, nuclear waste, spent fuel, and development of the fast breeder reactor remained unresolved. The future of nuclear power was in doubt, and even its most vigorous proponents were pessimistic.

HYDROELECTRIC POWER

Hydroelectric power received only very limited attention in the period following the oil embargo. Except in Alaska, there appeared to be few opportunities for high dams that might add significantly to the hydroelectric capacity of the nation. Most of the potential was tied up either with additions to existing high dams or with the movement to electricity generated from low dams.

Some federal money was expended on research and development aimed at hydroelectric power, and certain other subsidies were provided. For example, government loans were made available for up to 75 percent of the cost of small hydroelectric projects (Congressional Quarterly, Inc., 1981, p. 201). The goal of this program was to move the nation back to a position similar to that of a time when low dams had provided a major source of electricity, before they became economically uncompetitive.

In sum, policy actions taken with regard to hydroelectric facilities were of a very limited nature.

OIL SHALE

By 1974 the oil embargo had propelled interest in oil-shale development to a high level. In that year the Department of the Interior offered six leases—two each in Colorado, Wyoming, and Utah—under a prototype leasing program (*Oil and Gas Journal*, 1981a, p. 32). The intent of this program was to provide a basis for testing the engineering, economics, and environmental impacts of oil-shale development.

In the period immediately following the embargo the two Colorado leases sold for unexpectedly high bonus bids, one for more than $210 million and the other for more than $117 million. The two leases in Utah sold for more than $75 million and for more than $45 million. Under the terms of these leases two-fifths of the bonus price would be excused if that amount was invested in the development of the lease.

During the years following the issuance of the leases, interest in oil shale waxed and waned as the industry gained information and experience in the technical, economic, and environmental aspects of oil-shale production.

Following the Iranian revolution interest in oil shale experienced another peak. During 1980 the DOI undertook work on a proposal to issue four more leases under the prototype program and began planning for a permanent oil-shale leasing program. By the end of 1980 the federal government had issued four leases (*Oil and Gas Journal*, 1981). Work was under way on the engineering design and early construction for tests on two of those leases. In addition, initial work was under way on two oil-shale operations on private land.

The movement toward oil-shale development was reinforced by action of the Congress to support synthetic-fuels commercialization with the passage of the Synfuels Corporation legislation (Energy Security Act, 1980). By the end of 1980 most of the details had been worked out with the DOE for a contract that would provide a price guarantee to an oil-shale facility being developed on private land by the Union Oil Company. A parallel development was also worked out whereby federal loan guarantees of up to $1.2 billion would be provided to the Oil Shale Company as a part of its joint venture with Exxon. These developments were ultimately approved by the Reagan administration and were moved to the Synfuels Corporation (Corrigan, 1981, p. 430). By the end of 1980 it appeared that the first commercial-scale oil-shale test facilities in the United States would be in operation in the 1980s. It was generally believed that liquids from oil shale would be cheaper than those from synthetic-coal processes. Uncertainties remained, however, about the technology, economics, and environmental consequences of large-scale oil-shale developments.

Experience during the 1970s with regard to oil shale supported the view that only with federal assistance would the nation test a number of commercial-scale operations. Although hundreds of billions of barrels of liquids were to be had from oil-shale deposits, uncertainty remained about whether it was a viable economic activity for the private sector. The Synfuels Corporation appeared to ensure removal of that uncertainty.

TAR SANDS

Although interest in developing tar-sands deposits in Utah increased during the 1970s, legal problems stood in the way of development on federal lands, where most of the tar sands occur. Before these sands could be developed, it was necessary for the Congress to establish new federal leasing categories. In some of the tar-sands areas the lands were already held under oil-and-gas leases, but the terms of those leases did not permit tar-sands development.

In 1980 the DOI framed a program to allow tar-sands development. The favored approach was to establish a new lease category called "combined hydrocarbon leases." Under this category it would be possible for holders of oil-and-gas leases to develop tar sands as well. No formal action had taken place by the end of 1980, but such a bill was introduced in the Congress and signed by Reagan in 1981 (Combined Hydrocarbon Leasing Act, 1981).

GEOTHERMAL ENERGY

Geothermal power received a substantial amount of attention following the oil embargo, including accelerated leasing of federal geothermal lands, a very substantial federal R&D program, tax benefits, and a program of geothermal loan guarantees.

In 1970, in an effort to make federal geothermal lands available for development, the Congress had passed the Geothermal Steam Act. With the embargo the federal government accelerated its efforts in this area, and the first geothermal leases were issued in late 1974. To assure that both leasing and post-lease exploration and development would take place rapidly, the Department of the Interior established a special Geothermal Office to oversee activities on public lands.

Along with the leasing program, government support for geothermal R&D grew rapidly. Between 1974 and 1980 federal appropriations for geothermal energy rose from $19 million to $181 million (Executive Office of the President, 1976, p. 253; Executive Office of the President, 1982, p. 434). R&D activities ranged from the development of drilling and exploration technology to research on improved energy production from geothermal energy sources to the underwriting of small-scale demonstration plants.

To encourage development of commercial geothermal-energy facilities, the Congress also provided two other types of support in the period following the embargo. In 1976 the Congress established a loan-guarantee program for geothermal facilities (Congressional Quarterly, Inc., 1976b, pp. 153–59). Three years later, in 1979, geothermal energy projects were given tax advantages in the form of depletion allowances and the opportunity to write off intangible drilling costs (U.S. President, 1977b, p. 582). By the end of 1980, however, the only large-scale geothermal energy development was one that had been operational before the embargo, at The Geysers, in northern California.

ORGANIC WASTES

Organic wastes were universally seen as an attractive energy source (Science and Public Policy Program, 1975, p. 10–13). Federal R&D funds were provided to underwrite experimentation with and demonstration of organic-waste-fueled energy facilities, ranging from boilers in electric power plants to small-scale plants designed to convert wastes into liquid and gaseous fuels. Compared with the funding for other energy sources, however, the R&D dollars spent on organic wastes were quite limited. In 1976, for example, solar R&D received $290.4 million, while organic waste R&D received $4.3 million (Energy Information Center, 1976, p. 13).

The Congress did pass legislation aimed at encouraging the utilization of organic wastes in the form of price supports, loans, and loan guarantees. The legislation establishing the Synthetic Fuels Corporation also authorized financial assistance up to $250 million for organic-waste facilities. Although progress was made in this area, organic wastes represented but a miniscule part of the nation's energy supply by the end of 1980, and their potential was seen as very small.

SOLAR ENERGY

Policy for solar energy following the embargo had three characteristics. First, there was a marked increase in federal support for solar R&D. Between 1974 and 1980 federal expenditures for solar R&D grew from $45 million to $597 million (Rosenbaum, 1981, p. 16). Second, solar energy was almost universally approved—it was all but impossible to identify active opponents. Third, however, there was pervasive skepticism that solar energy would make a major contribution to the nation's energy needs in the short- to midterm future.

In October, 1974, the Congress created the Energy Research and Development Administration (ERDA) and specified that one of its six major program areas would focus on solar, geothermal, and advanced energy systems (Energy Reorganization Act, 1974). Solar R&D was in fact the major focus of that program area. In the same year the Congress passed the Solar Energy Research, Development, and Demonstration Act. This legislation established the Solar Energy Research Institute (SERI), which quickly became one of the key actors in the solar energy community. In fact, SERI, under the leadership of Dennis Hayes, rapidly became one of the most controversial participants in that entire energy policy community (Congressional Quarterly, Inc., 1982, pp. 67–70).

Beginning with these two pieces of legislation, support for solar research and development grew both rapidly and consistently through the 1970s. Proposed solar technologies supported by the federal government covered the spectrum from research on passive energy through active solar technologies, biomass,

Table 9.1. Public Acceptance of Nuclear Power

Energy alternatives	Chosen in top three preferred, percent		Top three minus least preferred, percent	
	Gallup survey*	Roper and Cantril survey†	Gallup survey*	Roper and Cantril survey†
Solar energy	66	61	63	55
Energy conservation	45	35	41	32
Synfuels	38	26	33	16
Hydroelectric power	34	31	26	21
Coal	36	36	23	27
Oil and natural gas	34	28	22	19
Nuclear energy	27	23	−18	−10

*Data from Gallup national survey of homeowners conducted for the Solar Energy Research Institute, October–November, 1980.

†Data from Roper and Cantril national survey of adults eighteen and older conducted for the Council on Environmental Quality, January–March, 1980.

Source: Reprinted with permission from Krieger 1980, p. 73. Copyright 1982 American Chemical Society.

and wind energy options to energy generated from ocean thermal gradients. By the end of 1980, however, none of these technologies had been demonstrated to be economically competitive with conventional energy sources.

Perhaps the most striking pattern with regard to federal solar R&D policy was the broad base of congressional support (reflected in public opinion, as shown in table 9.1). It was not uncommon during this period for the Congress to appropriate more money for solar R&D than the various administrations requested. This broad base of support for solar R&D was not, however, translated into a large-scale, well-funded federal effort aimed at the widespread utilization of solar energy options.

Throughout the later part of the 1970s and 1980 the Congress provided diverse tax credits and incentive programs aimed at accelerating the movement to solar energy. These multiple efforts never consolidated into a well-organized, heavily funded effort, however. Rather, in most instances the incentives were token additions to other pieces of energy legislation. A review of the solar initiatives suggests that adding a subsidy for solar energy was frequently a part of the coalition building necessary to pass energy legislation aimed at other options. Many solar advocates argued that only a major large-scale, federal solar effort would be successful in moving the nation to a heavier reliance on solar energy. Even President Carter's call for a 20 percent energy contribution from solar power by the year 2000 was not enough to trigger the large-scale program that was conceived as necessary (Congressional Quarterly, Inc., 1981, p. 71). In fact, the general consensus was that Carter's optimistic goal was unlikely to be achieved.

The largest single commitment of federal funding for solar energy came in 1980 with the passage of the synfuels legislation described under both "coal" and "oil shale" above. That legislation authorized $1.2 billion for loans, loan guarantees, price guarantees, and purchase agreements to underwrite and encourage the production of alcohol fuels from organic materials. The legislation also authorized a $525 million solar energy bank. Again, however, a review of the debate suggests that these items were added to the Energy Security Act, at least in part, to build the coalition necessary to pass the synfuels legislation. Certainly there was no expectation that the same kind of emphasis would be placed on solar energy that was being placed on synfuels from coal and oil shale.

By the end of 1980, then, solar energy remained a hope for nearly everyone, but there was no expectation reflected in presidential-congressional decisions that it would play a major role in meeting the short- to midterm energy needs of the nation.

ELECTRICITY

Issues associated with electric power were central to the energy policy debate that took place in the years following the oil embargo. As discussed earlier, one key issue concerned moving electric power generation away from oil and gas to other fuel sources, specifically coal and nuclear power. Another continuing issue was the effort to slow the growth in consumption of electric power and, within that, to manage the pattern of electricity consumption.

Two patterns characterized the electric power industry before 1973. First, the industry generally priced electric power on a declining block rate. Simply stated, the larger the quantity of electricity used, the cheaper the price became. In a time of energy shortage this pricing pattern seemed to encourage profligate use rather than conservation. Second, demand for electricity varies greatly over any twenty-four-hour period and, for most utilities, seasonally over the course of the year. Generally, demand is high during any given day in the morning immediately before the departure of people for work and in the evening immediately after their return from work. In areas of the country where the temperature is very hot during the summer, there is a much higher demand for power for air conditioning than is the case during the winter months. A reverse peak occurs in those parts of the country with very cold winters.

Both the utilities and the state regulatory agencies had traditionally followed policies that ensured their capacity to meet these peak demands. Peak-demand electricity was usually produced by generators driven by either turbine or internal-combustion engines. This peak-demand-generating equipment had advantages in that it could be started and shut down very quickly and was relatively low in capital cost. On the other hand, turbine and internal-combustion generators were heavy users of scarce gas and oil. Therefore, one element of policy for electric power was aimed at reducing the need for peak-demand electricity.

Efforts to conserve electricity focused on a restructuring of the electric utilities' pricing patterns, which came to be known as "utility rate reform." As part of his comprehensive energy proposals in 1977, President Carter called for comprehensive utility rate reform and a major role for the federal government in achieving that reform (U.S. President, 1977b, p. 582). Carter proposed eliminating the declining-block-rate pricing pattern and charging significantly higher prices for energy used during peak-demand periods.

The Carter proposals brought vigorous opposition from industrial users, who argued that they deserved the financial advantages of lower rates because it was much cheaper to provide them with electricity than residential and commercial customers. The rationale behind industry's position was that it did not require a large investment in a delivery and distribution system. Industry was also opposed to a rate structure that accelerated the cost of electricity during peak periods, arguing that its power requirements were stable and did not create the peaking problem. Moreover, industry's requirements were much less flexible than those of residential and commercial consumers. The key issue, however, revolved around a set of federal initiatives that would have taken control of a large measure of the regulatory responsibility then resting in state utility regulatory agencies (Congressional Quarterly, Inc., 1981, p. 62).

In 1978 the Congress acted on legislation dealing with utility rate reform; however, the legislation was so watered down that its primary effect was to place the federal government in the position of intervener in state rate-setting activities and an external prod to the state agencies for rate reform (Public Utility Regulatory Policies Act, 1978).

In fact, a large number of the state regulatory agencies did modify the pattern of utility pricing. In few cases, however, did the changes go as far as Carter's original proposals called for. By the end of 1980 the role of the federal government in establishing rates for electric utilities was quite limited.

ENERGY CONSERVATION

Between 1973 and the end of 1980 energy conservation became a broadly supported objective. Before the embargo conservation had been described as a moral imperative and, conversely, as a threat to the nation's economic well-being (Congressional Quarterly, Inc., 1981, p. 55). By the end of 1980 the debate over conservation had moved away from these rhetorical extremes, and a broad range of policy actions had been taken to encourage or enforce conservation. Central to the development of a consensus supporting energy conservation was the demonstration that economic growth could occur without a parallel pattern of energy growth. The amount of energy needed to generate a dollar's worth of economic activity declined in every year but one between 1973 and 1980, and this pattern of decline was expected to continue (Congressional Quarterly, Inc., 1981, p. 54).

The key factor behind the striking improvement in energy efficiency and, therefore energy conservation, appears to have been the rising price of energy. In addition, however, the president and the Congress had taken a number of steps aimed at establishing performance standards for energy, providing subsidies and tax incentives for energy-conservation activities, educating Americans about the advantages of energy conservation, and appropriating large amounts of money for R&D aimed at energy conservation.

Perhaps the symbolic turning point occurred with the election of Jimmy Carter as president. Energy conservation was a primary goal of the National Energy Program proposed by President Carter ninety days after his inauguration (U.S. President, 1977d, p. 566). Although Congress did not enact his energy plan, the objections to it did not proceed from opposition to conservation. The failure of the plan resulted from controversy over the complex set of instruments that the president proposed to use to achieve conservation.

The price of imported oil quadrupled at the time of the embargo and doubled again with the outbreak of the revolution in Iran. These price increases made obvious the economic advantages of energy conservation. The industrial sector of the economy responded quickly once it became clear that capital investments aimed at saving energy had a high payoff (Congressional Quarterly, Inc., 1981, p. 27).

By 1978 and 1979, when the Congress passed the Natural Gas Policy Act and President Carter began the deregulation of oil, the economic advantages of energy conservation had become even more widely perceived. The deregulation of oil and gas prices, which allowed domestic energy costs to move toward the OPEC price, added an economic sanction to profligate energy use. High-priced energy caused Americans to use energy more efficiently and to make some modifications in life-style. These modifications ranged from setting house thermostats lower in the winter and higher in the summer to driving less.

The incentives for energy conservation provided by high prices were supported by other public-policy actions. Energy conservation was a major component of several pieces of legislation passed between 1975 and the end of 1980.

Among the most immediate responses to the oil cutoff were efforts by government to require energy-efficient practices by the public. These efforts began in late 1973 with the passage by the Congress of a fifty-five-mile-an-hour speed limit on the highways and the establishment of twelve-month daylight saving time (Emergency Highway Energy Conservation Act, 1973; Emergency Daylight Saving Time Energy Conservation Act, 1973). The negative public reaction to these two acts was a harbinger of the difficulty that government would have with every effort to mandate energy conservation.

Unquestionably the most severe conservation mandate passed by the Congress was the requirement for auto efficiency standards in the Energy Policy and Conservation Act of 1975. Those standards required that passenger-car manufacturers move through a series of graduated steps to a point where the fleet average gas-mileage performance would be 27.5 miles per gallon in 1985. This action was vigorously opposed by the auto industry (Murphy, 1975, p. D17). The oil-saving potential of efficient cars, however, was so compelling that the Congress took the action in face of the opposition. In fact, the high price of gasoline produced such a massive movement by Americans to fuel-efficient autos that the miles-per-gallon standard was essentially moot by 1980. In response to buyer demand, General Motors was estimating that its fleet average would be 31 miles per gallon by 1985 (*Washington Post,* 1980, p. B1).

In 1975 and 1976 the Congress also took initial actions aimed at establishing energy-performance standards for major appliances and for buildings (Energy Policy and Conservation Act, 1975). The first step simply required appliance manufacturers to put labels on their products indicating energy performance. Two years later, in 1978, as a part of the National Energy Conservation Policy Act, Congress took another step by directing the preparation of energy performance standards for major appliances (National Energy Conservation Policy Act, 1978). These standards were to become mandatory in the mid-1980s. Standards for building construction were initially addressed by the Congress in 1976 as a part of the Energy Conservation and Production Act. Under that legislation the Department of Housing and Urban Development was to establish efficiency standards for all new commercial and residential buildings. The legislation specified that proposed building standards must be approved by the Congress before they became effective. Once the standards were approved, federal financial assistance for the construction of new buildings would be available only if the various states certified that the buildings met the standards.

The responsible federal agencies had great difficulty in establishing acceptable energy-performance standards both for major appliances and for buildings. There was broad opposition to the standards in the affected industries (U.S. Congress, 1977c, pp. 611–13, 516–17). The problems of formulating regulations that contained detailed requirements were the same kinds of problems the government faced in efforts to provide close regulation of the oil-and-gas industry. None of these mandated efficiency requirements were in place by the end of 1980.

During the postembargo period the government also established temperature standards for commercial and public buildings. Those standards specified that temperatures should be maintained at no more than 65 degrees in the winter or less than 78 degrees in the summer (U.S. President, 1973a, p. 1319). Although the government made no effort to enforce these standards, they were generally perceived to have a beneficial effect. The reason for their effectiveness was that building owners had an economic incentive to apply them. The building owner could say, "It's warmer than we'd like or cooler than we'd like because that's what the federal government requires," even though the owner's primary motivation was the economic advantages he received.

On balance, government efforts at establishing performance standards aimed at energy conservation probably played only a marginal role in the nation's success in moving toward greater energy conservation. These standards, however, did contribute to educating Americans that energy efficiency was a plus for their pocketbooks.

Other efforts to stimulate energy conservation included a program initiated in the Energy Policy and Conservation Act of 1975 whereby the federal government funded state energy-conservation programs. Funding for state programs was continued through the late 1970s. The overall effect of the programs is impossible to determine, but it clearly contributed to the growing awareness of Americans of the advantages of energy conservation.

Beginning in 1976 with the passage of the Energy Conservation and Production Act, the Congress began increasing federal subsidies for actions aimed at improved energy conservation. The Energy Conservation and Production Act provided grants to states, Indian tribes, and city governments for insulation and other weatherization investments. The subsidies ranged from federal funds for weatherizing the homes of low-income families to various subsidies and loan guarantees for broader programs of residential, commercial, and industrial energy conservation. In 1978 the Congress passed legislation providing a wide range of tax credits for conservation activities (Energy Tax Act, 1978). For example, the homeowner was allowed to write off 15 percent of the first $2,000 spent on energy-conserving improvements.

Many observers believe that one of the most effective actions taken by the Congress during this period was the requirement in the National Energy Conservation Policy Act of 1978 that electric and gas utilities undertake a concerted program to encourage their customers to conserve. Under this legislation utilities not only provided energy audits of buildings but also arranged financing for energy-conservation measures that could be paid for as part of the monthly utility bill. It is doubtless correct that one of the reasons for the success of this program was that many of the utilities began to see it as advantageous for them to stimulate conservation by their customers. The high capital cost of new electric generating plants, for example, made it economically advantageous for electric utilities to stimulate conservation.

Federal research and development appropriations for work on energy conservation grew from $105 million in 1974 to $568 million in 1980 (Executive Office of the President, 1975, p. 57; Executive Office of the President, 1982, p. 135). As with solar R&D, there was a broad base of support for conservation-oriented R&D in the Congress. It was common throughout the period from 1973 to 1980 for the Congress to appropriate more money for conservation R&D than the various administrations requested.

In sum, by 1980 there was a broad consensus in support of energy conservation. Conservation was a continuing focus of public policy during this period. Although there were no massive conservation efforts by the federal government, the combination of large subsidies with information programs tied to the increasing cost of energy led to striking results. Alternatively, use of sanctions such as performance standards to promote conservation generally was not in favor by 1980. By the end of 1980 conservation had provided more new energy to the nation than any other source.

CHAPTER 10

Putting the Pieces Together

The preceding chapters have investigated the evolution of presidential-congressional decision making between 1973 and 1980 in the twelve energy resource areas. In a concurrent decision-making process the search was under way for structures and organizations capable of stable policymaking and management under the changed circumstances of energy. In other words, a new energy policy community was being sought.

It should be recalled that physical policy communities are characteristically both complex and informal. Actors and interests participate in policy systems for one or a combination of three motives: performance, impacts, and power. Before the oil embargo scientific-technical energy participants had been primarily motivated by a concern with performance defined in terms of engineering and economic considerations. The makeup of the scientific-technical group expanded to include actors and interests with expertise in all the twelve resource options. Second, the scientific-technical participants expanded to include professionals concerned with environment and health.

A similar expansion in the participants motivated by impact concerns occurred. Before the 1970s the dominant impact concerns were economic and legal. Traditionally economic impacts were of primary interest to producers of energy, and legal impacts were the concern of government organizations and agencies. When protection of the environment became a major area of concern, the number of impact-motivated participants in the energy policy system substantially increased to include government agencies with responsibilities for environmental-protection control and private-sector interest groups that were primarily concerned with the environmental and health impacts of energy development.

The period from 1973 to 1980 was characterized by vigorous efforts by congressional committees, executive agencies, state governments, and many of other actors and interests to gain control over energy. Energy was seen as an area of great national importance, and a role in controlling it offered power.

To structure a stable energy community, the president and Congress had to determine the participants and the roles they would play. The following sections investigate the decisions that determined who the participants would be in each of four organizational settings: (1) the federal government, (2) state governments, (3) the private sector, and (4) nonprofit research organizations.

THE FEDERAL GOVERNMENT

The first initiatives within the federal government to organize for a comprehensive energy approach as opposed to a fuel-specific approach occurred in the early 1970s, before the embargo. Senate Resolution 45, passed in 1970, authorized a major study of energy under the direction of Senator Henry Jackson of Washington. That study was to make Jackson a preeminent member of Congress in the field of energy during the early years of the energy crisis. In 1971, as a result of a comprehensive study of the federal executive structure President Nixon recommended the creation of a new Department of Energy and Natural Resources (DENR; U.S. President, 1978). Both Jackson's study and the DENR proposal were indications of a growing perception that the nation needed to take a more comprehensive approach to energy and organize to formulate policy for energy rather than for specific fuels. During 1970 a federal agency was created that would play a very important role in the evolution of the energy policy community. That organization was the Environmental Protection Agency (EPA; U.S. Congress, 1970, p. 1). The EPA was created soon after the passage of the National Environmental Policy Act, and most of the government's environmental regulation programs were consolidated in that agency.

From its inception the EPA addressed energy as a major area of concern. In the early period it had charge of implementing the Clean Air Act of 1970, and particular attention was focused on air emissions from electric power plants and other large boilers. Before the embargo EPA had undertaken major efforts both to establish emission standards and to stimulate the development of cleanup technology for coal-fired power plants.

240

From hindsight it is clear that the establishment of the EPA ensured that the federal government would be deeply involved in environmental aspects of the energy policy struggles that would follow the embargo.

In the months immediately preceding the embargo perception of the need for a consolidated federal overview of energy led to a number of initiatives in the White House (Executive Orders 11712, 11703). For example, the Domestic Council, under presidential assistant John Erlichmann, established a Subcommittee on Energy and an Oil Policy Committee. In February, 1973, President Nixon established a Special Energy Committee and a National Energy Office in the White House (Congressional Quarterly, Inc., 1981, p. 135). In June he appointed John Love, governor of Colorado, to head a new Energy Policy Office in the White House. All these organizational actions reflected a growing awareness that the nation's approach to energy was inappropriate for dealing with the changing facts of energy. In December, 1973, following the embargo, President Nixon established a new Federal Energy Office in the White House and appointed William Simon as its head. The Federal Energy Office was self-consciously temporary and remained in existence only until Congress established a new agency, the Federal Energy Administration (FEA), in May, 1974. The FEA, created with a two-year life span, was assigned the responsibility of managing shortages and allocating energy supplies, as well as carrying out a number of other information-collection and regulatory activities.

In October, 1974, the Congress passed the Energy Reorganization Act, which dissolved the Atomic Energy Commission and established two new agencies: the Energy Research and Development Administration (ERDA) and the Nuclear Regulatory Commission (NRC; Energy Reorganization Act, 1974).

In the ERDA were consolidated all federal energy research and development. The rationale behind the creation of the new agency was twofold. First, it reflected the perceived need to provide unified federal management of energy research and development. Second, it reflected a desire to put considerable emphasis on sources of energy other than nuclear power, which up to that time had been receiving most of the federal energy R&D support. The creation of the NRC was also in part a

241

response to a widely held view that the Atomic Energy Commission had sacrificed safety and environmental protection to the promotion of nuclear power.

The Nixon administration supported the creation of the FEA, the ERDA, and the NRC, but taken together, this set of organizational arrangements represented a compromise on the proposed Department of Energy and Natural Resources. The regulatory policy role of the FEA and the R&D role of ERDA were split, whereas under the DENR concept the activities would have been consolidated in a single agency.

The next major organizational initiative in the executive branch was the creation in 1977 of the Office of Surface Mining (OSM) in the Department of the Interior to implement provisions of the new Surface Mining Control and Reclamation Act. Although the OSM was organizationally only a bureau within the department, it represented an important indication of the evolution of the policy process. To paraphrase the comments of one observer: If the Congress is only mildly serious about an issue, it passes a law. If it is serious, it passes a law and appropriates money. If it is very serious, it passes a law, appropriates money, and creates a new organization for the specific purpose of carrying out the legislation. The intensity of congressional commitment to reclamation was reflected in the creation of the Office of Surface Mining.

The next major organizational modification in the federal executive was the creation, also in 1977, of the Department of Energy (DOE; Department of Energy Organization Act, 1977). In the DOE were consolidated most federal energy-related activities. The most important agencies brought together with creation of the new Department of Energy were the Federal Energy Administration, the Energy Research and Development Administration, and the Federal Power Commission. In addition the DOE took over certain responsibilities from other departments, specifically Interior, Housing and Urban Development, Transportation, Navy, Commerce, and the Interstate Commerce Commission. Unlike Nixon's proposal for a Department of Energy and Natural Resources, however, the enabling legislation for the DOE left responsibility for the management of energy on federally controlled lands in the Department of the Interior. Similarly, the Office of Surface Mining remained in the Department of the Interior.

The final federal action in the effort to formulate a national energy policy and construct a stable energy policy system was the creation in 1980 of the Synthetic Fuels Corporation (Energy Security Act, 1980). The corporation was established with the legislatively mandated purpose of moving the nation rapidly toward development of a major commercial synthetic-fuels industry.

Thus by the end of 1980 the executive branch had five major organizational components concerned in part or entirely with energy policymaking: the Department of the Interior, the Nuclear Regulatory Commission, the Environmental Protection Agency, the Department of Energy, and the Synthetic Fuels Corporation. Several characteristics of this constellation of federal agencies warrant specific comment. Perhaps the most striking is that four of the five organizations were either totally or in substantial part devoted to understanding and managing the environmental and health and safety impacts of energy activities. The DOE had primary responsibility for developing an energy policy and promoting the development of an energy system that could meet the nation's needs. The DOE's enabling legislation had also, however, authorized establishment of a specific component of the agency whose concern was with the environment. Similarly, the Department of the Interior had primary responsibility for promoting energy development on federally controlled lands, but all of its energy-related bureaus were responsible for formulating and enforcing environmental regulations. Indeed, for one of those bureaus, the Office of Surface Mining, environmental protection was its primary reason for being. Of the five agencies only the Synthetic Fuels Corporation was without a major environmental responsibility.

The EPA had overall national responsibility for environmental protection (U.S. President, 1971, p. 545). A significant portion of its activities was energy-related, and it was to be a consistent and continuing participant in the formulation and implementation of national energy policy. The Nuclear Regulatory Commission existed primarily to ensure that nuclear energy was developed in a safe and environmentally acceptable manner (Energy Reorganization Act of 1974).

In sum, the federal organizations that were to be continuing participants in the energy policy system represented a full

243

range of energy-related concerns. Their mission was to oversee not only the scientific and technical development of energy resources and conservation but also the scientific and technical development necessary to avoid or manage adverse environmental and safety impacts. Thus in these agencies actors and interests concerned with the impacts of energy development had representation at the federal executive level. Energy producers had access to the Department of Energy, the Synfuels Corporation, and the Department of the Interior. For environmental interest groups the EPA, the Nuclear Regulatory Commission, and various components of the Department of Energy and the Department of the Interior provided access points. In sum, by the end of 1980 all segments of the energy policy community—state governments, the private sector, and the nonprofit research organizations—had access to the federal government.

By contrast, in Congress the organizational modification resulting from the energy crisis was much less extensive. Two primary committee changes were made during this period. One was the abolition in 1977 of the Joint Committee on Atomic Energy (Congressional Quarterly, Inc., 1981, p. 180). The joint committee had been created in 1946, when Congress moved responsibility for nuclear weapons from the Army Corps of Engineers to the new civilian Atomic Energy Commission. The joint committee was unique in that it was the first and only permanent joint committee with authorization responsibility in Congress. Both its critics and its admirers agree that it played a significant role in moving the nation rapidly in the direction of commercial nuclear power. When the Atomic Energy Commission was dissolved in 1974, however, the committee went into precipitous decline and was dissolved three years later. It should be noted that two of the criticisms of the committee that in the end doubtless contributed to its dissolution were (1) that it had sacrificed nuclear safety to the promotion of nuclear power and (2) that the existence of such a powerful committee in the Congress had been one reason the United States had not devoted more effort to the development of energy sources other than nuclear power. With the abolition of the joint committee responsibility for nuclear power was fragmented in the Congress.

The second major Congressional change with regard to energy

was the metamorphosis of the Senate Committee on the Interior into the Senate Committee on Energy and Natural Resources. The broadened jurisdiction of the new committee made it one of the dominant centers of energy-related activity in the Senate. The other significant center was the Senate Committee on Environment and Public Works. Like the Environmental Protection Agency, the Committee on Environment and Public Works gained its responsibility and control over energy through its broad authority over the environment (Congressional Quarterly, Inc., 1976a, p. 3239). In the House of Representatives there was not the degree of concentration that occurred in the Senate. Responsibility was divided among three committees: Commerce, Interior, and Science and Technology.

The structure of congressional participation in energy policy-making provided the same kind of access to Congress for the scientific, technical, and impact interests that the new agencies provided in the executive branch. The fragmentation of responsibility for energy among the committees of the House and Senate meant that there were avenues to Congress for energy-concerned actors and interests in state governments, the private sector, and the nonprofit research components of the energy policy community. This representation ensured federal support for research and development on all energy-resource options, as well as sensitivity to the full range of economic, legal, and environmental impacts of energy-related activities.

STATE GOVERNMENTS

The capacity of state governments to participate in a new energy policy community varied greatly across the United States. The state governors had been brought into the energy policy process at the time of the oil embargo as a result of federal efforts to establish and manage an allocation program for oil. The initial allocation program worked out by William Simon when he was the head of the Federal Energy Office of the White House had placed heavy emphasis on participation by the state governors. With the end of the embargo the states responded in a variety of ways. There were, however, two areas in which federal programs underwrote the establishment of state organizations that were to play a continuing role in en-

245

ergy activities. The first involved environmental and mining-reclamation regulations. The general pattern was that the federal government established minimum standards, approved the states' approaches, and provided funding for the states to enforce the regulations. A similar approach was followed in the effort to stimulate energy conservation through federally supported, state-managed energy conservation programs, beginning with state energy-conservation plans and expanding to programs aimed at weatherization of low-cost housing and other conservation-related activities.

Some states, such as New York and California, established their own energy R&D organizations and levied taxes to help support them, the rest of the money coming from grants or contracts with the Department of Energy. In coastal states and in western states with important federal lands, state governments asserted the right to participate in the policy process formulated for the development of federal lands. Generally state participation was heaviest at the leasing stage.

In sum, although the pattern was not consistent, by the end of 1980 there was a diverse capability in various states to participate in energy policymaking. In certain areas such as conservation, protection of the environment, and management of federal lands, arrangements were in place for continuing participation by the states in energy policymaking.

THE PRIVATE SECTOR

By 1980 the major private-sector participants in energy policymaking could be categorized in three groups: energy producers, environmental interest groups, and private-sector profit-making R&D organizations. With regard to energy production, the major actors were the same ones that had participated before 1973. The most important of these were the oil companies, which had expanded horizontally over the period 1973 to 1980 to become energy companies. Many of the large oil companies had expanded into coal; many had undertaken oil-shale activities. Some companies had developed solar and geothermal capabilities. And most of the energy companies were experimenting with a wide range of synthetic-fuel options.

The Interstate Gas Transmission Companies remained the managers of the interstate gas pipeline system, and several of them had moved into oil and gas development as well as undertaking experimental or, in one instance, commercial-scale coal synfuels programs. The electric power industry had expanded its research and development activities with the creation of the Electric Power Research Institute in 1973. Electric utilities were also promoting energy conservation both through energy audits and financing of weatherization of homes and through the use of new technologies that enabled them to manage more efficiently peak-period demand. The nuclear power industry remained a major participant in energy policymaking, although so many uncertainties developed in the nuclear power field that by the end of 1980 the influence of the nuclear industry was in serious decline. The influence of the coal industry had grown during this period, though the role of coal had not become as important as either the industry or several presidents had hoped.

In the second category of major private-sector participants were the environmental interest groups, which had demonstrated both the willingness and the capacity to participate in energy policymaking on a continuing basis. The Audubon Society, the Sierra Club, the Natural Resources Defense Council, and other national and regional organizations were consistent participants in energy policymaking during this period and showed every evidence of being continuing participants in an energy policy community. These organizations represented major constituencies for both the environmentally oriented federal and state agencies and the committees of Congress with environmental responsibilities.

Finally, the rapid increase in federal expenditures on research and development in energy production and conservation and in the environmental impacts of energy facilities produced a large number of private-sector profit-making organizations that did everything from R&D to the preparation of environmental impact statements. Such large private-sector corporations as TRW had major organizational components devoted to energy R&D and studies for the various federal agencies with energy responsibilities. Added to these groups were various organizations assisting the energy industry in environ-

247

mental management by producing such products as equipment for flue-gas desulfurization plants, equipment to monitor the residuals from energy facilities, and models of atmospheric diffusion of emissions from coal-fired power plants to predict where pollutants would be carried by the wind.

Perhaps the most striking feature of the private sector by the end of 1980 was what did *not* occur. As early as 1973 Senator Adlai E. Stevenson III of Illinois proposed that the federal government establish its own oil-and-gas corporation (*Oil and Gas Journal,* 1973b, p. 90). The idea underlying what came to be known as FOGCO was that the federal government should have the same yardstick capability (independent source of information) for petroleum that the Tennessee Valley Authority provided for electric power. The proposal, however, never generated wide support. During this period many proposals were put forth for both vertical and horizontal divestiture of the large energy companies. In the oil industry vertical divestiture would mean division of the large companies into smaller production, refinery, distribution, and wholesale and retail companies. Although there were some close votes in the Congress on this issue, no legislation generated the necessary support to achieve passage into law (Congressional Quarterly, Inc., 1981, p. 175). The same fate awaited horizontal-divestiture proposals, which would force large energy producers to divide into smaller companies to ensure that no one company could be in the business of producing oil and gas as well as other energy sources such as coal and oil shale. Similarly, new policy initiatives failed to bring forth large new energy-production industries. Arguments by solar power advocates, for example, for a large federal program to promote a solar industry failed to receive congressional approval. Although many small companies were organized to develop solar technologies and produce solar power, the industry was a pygmy among the giants of traditional energy production.

Thus by the end of the 1970s it was apparent that the president and the Congress had decided that the major energy producers would continue to be the large companies of the years before the oil embargo. For the foreseeable future the primary sources of energy were likely to be fossil fuels, and those fuels would be supplied by large, established, integrated energy companies.

NONPROFIT RESEARCH ORGANIZATIONS

The availability of large amounts of federal monies for energy R&D and energy-related environmental R&D produced a marked expansion of activities in the nonprofit research organizations, specifically the universities, between 1973 and 1980. Nearly every discipline in higher education, from the social sciences through the natural sciences to engineering, concentrated on energy-related R&D. By the end of 1980 the federal government was supporting R&D aimed at accelerating the development of energy from all twelve of the resource options, as well as managing the environmental impacts of all those resources. Activities in the potential and theoretically possible categories were filled with an almost infinite range of optional technologies. Although the research community remained fragmented, working independently in the various resources, there was a much expanded and clearly identifiable scientific-technical component in the energy policy community.

SUMMARY

By the end of 1980 there had evolved in the United States a rudimentary energy policy community. That community had reached general agreement on the boundaries of the energy sector and general agreement on which of the energy sources fell into the available, potential, and theoretically possible subsectors. Further, there was general agreement on who the participants in the energy policy community would be. By the end of 1980, then, the boundaries had been established for an energy policy community with the potential for exercising stable energy management.

CONCLUSIONS

Part two opened by noting that following the embargo successive presidents and Congresses were faced with resolving four issues: defining energy goals, determining the sources of energy, choosing policy instruments, and identifying the participants in the energy policy community. The establishment of a stable, functioning policy system depended on a resolution of these issues. The achievements in these areas are summarized below.

Energy Goals

At the end of 1980 energy abundance remained a primary goal. "Abundance," however, had been redefined between 1973 and 1980. Before the embargo abundance had meant that surplus drove demand. The redefinition of abundance was reflected in new approaches to energy conservation. One approach was that Americans should change their life-styles. By the end of 1980 efforts to mandate life-style changes by legislation had few supporters. The other approach to conservation was improved efficiency in the use of energy. Growing support for improved efficiency characterized the period from 1973 through 1980. In fact, conservation, defined as efficiency, came to be perceived as a *source* of energy. By the end of 1980 energy abundance meant sufficient energy supplies to ensure no constraint on economic growth. In other words, after consumers had made all the efforts they believed economically advisable to conserve energy, there should be sufficient energy to meet the demand. Abundance, defined in those terms, was a high-priority national goal at the end of 1980.

The most pronounced change in the nation's energy goals had to do with cheapness. Before the oil embargo cheapness had meant that the per unit price of energy should in absolute terms decline over time. With the passage of the Natural Gas Policy Act and the deregulation of oil the cost of energy accelerated rapidly. Between 1973 and 1980 the price of oil went up almost tenfold (*Washington Post,* 1981, p. A25). Initial efforts to regulate prices were rejected, and a consensus developed that the national interest was best served by allowing energy to increase to what economists called its replacement cost. Simply stated, the nation came to abandon cheap energy as a high-priority national goal. This was the most significant presidential-congressional decision of the period from 1973 to 1980. It served as a powerful impetus for energy conservation and the search for new sources of energy.

Perhaps the most striking feature of the nation's struggle to define its energy goals is the tenacity with which three presidents and four Congresses held to the commitment to clean energy. Clearly, that commitment was an impediment to the more rapid development of domestic energy sources, particularly coal. Examples of the importance attached to environmental protection abound. Air-quality standards were more stringent in 1980 than they had been in 1973. During the 1970s stricter regulatory programs were adopted for reclamation of mined lands, for oil-and-gas development both on the OCS and on federal onshore lands, and for nuclear power. The commitment to clean energy pervaded both legislation and regulations. Further, broad-based federal organizations were put in place to enforce clean-energy requirements. The various Congresses and presidents demonstrated a continuing commitment by providing funds both for research and development on clean energy and for the enforcement of standards. The goal of clean energy thus ranked with abundance as a high priority. It came to mean that energy production should not violate a set of specific scientifically based effluent or emission standards or that the effluents and emissions should be as clean as the best available technology could make them.

Finally, with the imposition of the oil embargo the nation put secure energy on the national agenda as a high priority and kept it there from 1973 to 1980. The goal of security came to have three components: (1) adequate supplies of energy to allow the nation to meet its basic needs in a time of military conflict, (2) sufficient supplies of oil from domestic and reliable foreign sources so that energy could not be used as a political instrument to force changes in United States foreign policy, and (3) adequate energy supplies to assure the maintenance of economic stability. Stated differently, the United States should be protected against sudden denials of imports or sudden excessive changes in energy prices that would be damaging to the economy.

Concerns about security were both short and long term: one focused on protecting the nation against short-term denials of energy, and the other focused on developing the capability to supply the nation's energy needs over the long term as world oil supplies became insufficient to meet demand. To deal with the short-term problem, Congress established the Strategic Petroleum Reserve and passed standby rationing authority. The primary long-term responses to the security needs were the efforts to encourage conservation, to move the electric utility industry to the use of coal, to develop a major synfuels industry, and to support research and development.

By the end of 1980, then, abundance, cleanness, and security remained high-priority national goals, while the goal of cheapness was significantly downgraded in importance.

Sources of Energy

It will be recalled that immediately following the embargo the perception was that the domestic supply of each of the traditional fuel sources was or would be limited. Six new domestic energy sources were proposed as possible replacements: oil shale, tar sands, organic wastes, geothermal energy, solar energy, and conservation. Not all of those sources, however, like synthetic fuels from coal, or new technologies for generating electricity were commercially available. Most fell in the potential subsector.

By the end of 1980 the federal government had made a number of choices concerning the short- to mid-term sources of energy. First, energy conservation, defined as efficiency, came to be viewed as a source of energy, not a moral imperative or a threat to the nation. It was expected that conservation would continue to make available increasing amounts of energy over the long term. It was expected that, as the price of energy increased, all the consumption sectors would undertake energy-efficiency improvements. The quantity of energy needed to supply a dollar's worth of economic activity had declined every year but one following 1973, and that pattern was expected to continue.

The rising price of energy was also expected to sustain a continuing search for new domestic sources of oil and gas. In fact, by 1980 increased oil-and-gas drilling had momentarily halted the decline in reserves. Although most experts estimated that the halt in the decline in oil reserves would be brief, they expressed optimism that gas reserves would grow for several years.

Congress enacted legislation prohibiting the construction of new large boilers that used oil and gas, and coal was expected to be the primary boiler fuel for the short to mid-term. In fact, coal production increased by 40 percent between 1973 and 1980.

All three presidents sought to encourage the continuing construction of conventional light-water nuclear reactors. The rapid movement to the development of a breeder reactor, however, was inhibited by the opposition of President Carter. Projections about the future role of nuclear power were substantially less optimistic in 1980 than they had been in 1973.

Finally, the presidents and the Congresses made a fundamental commitment to the development of synthetic fuels, primarily from coal and oil shale. Legislation established the Synthetic Fuels Corporation with a daily production goal of two million barrels of oil-equivalent synfuels by 1992 (Energy Security Act, 1980). The synfuels legislation represented the only presidential-congressional decision during this period to provide massive federal support for moving an energy source from the potential subsector into the available subsector.

In sum, then, the primary sources of energy for the short- to mid-term future were to be conservation, oil and gas, coal, conventional nuclear power, and synthetic fuels.

In addition, the presidents and Congresses established an energy R&D program totaling approximately $5 billion. That program underwrote work in each of the energy resources. It was expected that these expenditures would over time develop economically competitive new sources of energy, such as solar power. Proposals for major policy initiatives to commercialize solar power, however, were rejected.

Policy Instruments

Between 1973 and 1980 the federal government experimented with various policy instruments aimed at stimulating conservation and developing alternative energy sources. Clearly the most important and effective instrument was deregulation, which allowed energy prices to rise to replacement-cost levels. By general agreement the deregulation of oil and gas proved to be the most effective instrument both to promote conservation and to accelerate the search for alternative sources of energy.

Some experiments were made with sanctions to stimulate conservation and the development of new energy sources. Sanctions ranged from establishing performance standards to taxes aimed at stimulating more appropriate or efficient use of energy.

By 1980, however, sanctions had largely been rejected except for regulations aimed at environmental protection.

On the other hand, a number of programs used various inducements to deal with the nation's energy needs, including subsidies and large direct federal expenditures. Significant examples were tax subsidies for weatherization of homes, the massive federal program to underwrite R&D, and the program to establish a commercial synfuels industry.

Participants in the Energy Policy Community

Most of the continuing participants in the energy policy community had been sorted out by the end of 1980. Perhaps the most significant development was that the large number of intermittent participants (for example, truckers seeking to protect low fuel prices) that sought to influence policy immediately after the embargo had dropped out of the energy policy process by the end of 1980. The primary reason for their departure was the decision to deregulate oil and gas. Once that decision was made and implemented, the many participants who had been motivated by the uncertain economic consequences of rapidly rising energy prices had either won or lost their struggle. Once the president and the Congress laid down the replacement-cost rule, the motivation for participating was eliminated for many actors and interests. Finally, the consistent pattern of congressional-executive action that assured participation by environmentally motivated actors and interests helped bring stability to the new energy policy community.

At the federal level five executive agencies were continuing participants: the Department of Energy, the Department of the Interior, the Environmental Protection Agency, the Nuclear Regulatory Commission, and the Synthetic Fuels Corporation. Similarly, the five committees of Congress with energy as a primary concern had established at least minimum ground rules concerning their jurisdictions. In the Senate the two committees were the Committee on Energy and Natural Resources and the Committee on Environment and Public Works. In the House the primary committees were those on Commerce, Interior, and Science and Technology.

From the state level came a diverse group of participants. The linking of the states into the energy policy system had generally stabilized by the end of 1980. In the private sector the major participants included the energy-producing companies, the environmental interest groups, and a number of R&D organizations. In the nonprofit research organizations scientists and engineers had established reasonably stable links with various federal funding agencies through R&D grants and contracts, and a pattern of reliable funding support had been established.

In sum, by 1980 the nation had constructed an energy policy system for the twelve sources of energy. Agreement had been reached on the participants in the energy policy community, and a rudimentary agreement had been reached by those participants on the boundaries of the energy policy sector and the activities that fell within each of the subsectors. The goals of national energy policy had also generally been agreed on. Finally, basic agreement had been reached on the policy instruments to be used in seeking to manage the nation's energy sources.

Congress had played the key role in establishing the consensus on energy policy. By 1980 the concerted efforts of three successive presidents and four Congresses to formulate an energy policy and construct an energy policy system had produced an energy policy system that held hope of successful energy management into the future.

PART THREE

The Arrival of the Radicals, 1980

CHAPTER 11

Energy Policy After 1980

As a result of [Reagan's energy] policies, perhaps the only Washington figures more pitiable than the energy lawyers, unemployed in droves, are the energy journalists, now frequently reduced to covering plane crashes and other real-world events rather than reporting administrative shadow-boxing with the ever-elusive "crisis."—Danny J. Boggs, White House energy adviser, May 1, 1982

Oil imports have gone down, but not necessarily because of anything the administration has done—unless they want to take credit for the recession. . . . They are too blasé about the potential for things going wrong in the world oil market.
—Philip R. Sharp, Representative from Indiana, May 29, 1982

The landslide victory of Ronald Reagan in 1980 brought to the White House a president who during his acceptance speech at the Republican National Convention had argued that the United States

must get to work producing more energy. . . . Large amounts of oil and natural gas lie beneath our land and off our shores. . . . Coal offers great potential. So does nuclear energy produced under rigorous safety standards. . . . It must not be thwarted by a tiny minority opposed to economic growth which often finds friendly ears in regulatory agencies for its obstructionist campaigns. (Congressional Quarterly, Inc., 1982a, p. 3)

Implicit in candidate Reagan's comments were two assumptions. First, the geological structures of the United States held huge quantities of conventional energy resources yet to be found and produced. Second, those resources had not been found and extracted because of counterproductive federal policies. Starting from these assumptions, the Reagan administration committed itself to reversing a trend toward government involvement in energy that had been growing since the oil embargo.

The Reagan administration saw energy from an economic perspective and rejected the view that energy required special policy attention. Energy was seen as no different from any other commodity. Reagan's answer to the energy problem was to move decision making into the economic marketplace. In its only general statement on energy, prepared in response to a congressional mandate for a biennial energy report, the Reagan administration characterized its approach to energy as follows: "This approach represents a radical departure from the prevailing policy instituted after the first shock of rapid oil price increases in 1973 and 1974" (Bureau of National Affairs, 1981b, p. 1107). As the Reagan people saw it, their reversal of energy policy flowed naturally from their economic philosophy:

The Administration's reformulation of policies affecting energy is part of the President's comprehensive program for economic recovery which includes elimination of excessive federal spending and taxes, regulatory relief, and sound monetary policy. When fully implemented, the economic recovery program will release the strength of the private sector and insure a vigorous economic climate in which the nation's problems, including energy problems, will be solved. (Bureau of National Affairs, 1981b, p. 1108)

If government was seen as the real culprit in the energy crisis, and the marketplace the panacea, energy policy was simple. The administration's first secretary of energy, James Edwards, put it bluntly. In his words the Reagan approach was designed to "produce, produce, produce" (Norman, 1981b). Both Reagan's rhetoric and his deeds have been remarkably consistent with this outlook. The response has been actions that assume steady increases in energy supply from conventional (fossil and nuclear) sources. Alternatively, the search for renewable resources and the focus on conservation have been downplayed. This change in approach disturbed even some Republican leaders. For example, William Cohen, former senator from Maine, termed Reagan's approach "energy Darwinism" because he saw it as lacking in concern for the public interest (Corrigan, 1982). Reagan's stance marked a dramatic departure from the energy policies of the three previous administrations. This chapter delineates the character of Reagan's break with the past, the policy implications of this new direction, and the consequences for the rudimentary energy policy system that existed by 1980. First, however, it is necessary to summarize briefly the major changes instituted by the Reagan administration. The starting point must be to understand the administration's approach to national energy policy goals.

THE NEW ENERGY POLICY GOALS

When the Reagan administration entered office, it inherited a set of energy policy goals around which a minimal consensus had evolved over the previous seven years. Those goals were abundance, cleanness, and security. Previous presidents and Congresses had clearly shared the view that specific government programs were necessary to achieve these goals. The Reagan administration differed from previous administrations in two ways, first, in the relative priority it gave to each goal, and, second, in the manner in which it believed the goals should be pursued. Administration spokesmen argued that the marketplace should be allowed to define both the content of the goals and the manner in which they should be achieved. Supply and demand, mediated by price, should determine the nation's energy future. Within this overall market philosophy, administration spokesmen repeatedly emphasized their belief that previous policy had impeded the achievement of the goals and that the market would more effectively contribute to their achievement.

The administration stated its position as follows:

The one thing that is certain about the future is that the exact path of energy development and markets is uncertain. Technological innovations, geologic discoveries, changes in the economy at home or abroad, political or military conflict, variations in public attitude—all of these are inherently unpredictable events that can alter the nation's energy situation drastically. Under the free market philosophy of this administration, the American people themselves will actually conduct a continuing national plebiscite in the marketplace to express their individual and collective evaluation of possible courses of action. Their actions will determine ultimately whether energy consumption per capita in this country rises or falls between now and the year 2000—and what our mix of energy sources will be at the turn of the century. (Bureau of National Affairs, 1981b, p. 1115)

Implicit in this orientation was a faith in the ability of the free market to reflect the nation's highly complex, pluralistic society through an economic choice mechanism. It was an outlook that, at its roots, was opposed to reliance on the political process. Thus there was resistance to setting goals for energy supply and demand, and opposition to activities that might be construed as planning (see Norman, 1981a). This philosophy made the marketplace itself the ultimate policy goal; supply and demand, mediated by price, should determine the nation's energy future. Not surprisingly, this outlook had drastic consequences for the three energy policy goals that had been determined by the end of 1980.

Abundance

With regard to abundance, President Reagan made it clear that his administration would emphasize the maximization of supplies rather than the reduction of demand. In general, the administration saw the previous attention to energy conservation as a major contributor to economic stagnation in the country except where conservation was motivated by market-determined price. The administration saw the policy focus on conservation as a reflection of a "no-growth mentality" and believed that abundance could be achieved only by reducing restrictions on the productive capabilities of the private sector. Typical of this perception was Secretary of Energy Edwards's comparison of the nation's energy system to human physiology: "The human body uses least energy when it's asleep or dead" (Congressional Quarterly, Inc., 1982a, p. 3).

Although the primary means of increasing energy production was seen as getting the federal government out of the private sector's way, there was, in the administration's view, one area in which the federal government had an important role: "The national government's most direct impact on America's energy future arises from its position as the steward of 762

262

million acres of publicly owned land. . . . The federal role in national energy production is to bring these resources into the energy marketplace." (Bureau of National Affairs, 1981b, p. 1107).

On only one energy issue, prices, did the administration find itself in agreement with previous actions. Reagan continuously expressed his firm belief that all energy prices should be deregulated. Indeed, the centerpiece of Reagan's energy policy was the decision to decontrol oil prices even more rapidly, and in 1983 he recommended the same for natural gas.

Cleanness

With regard to the goal of clean energy, the position of the Reagan administration was one of vigorous opposition to the actions of the federal government in the period from 1973 to 1980. Administration spokesmen declared that a major obstacle to solving the nation's energy problems was excessive environmental regulation. The aim of the administration was to establish a "reasonable balance" between energy and environmental values. In this area once again the new philosophy articulated a major role for the marketplace. According to the administration:

Energy-environmental conflicts arise because diverse groups place different values on energy itself and on individual elements of the existing environment which are altered as energy is produced, delivered, or used. Market mechanisms can resolve such conflicts if each of the resources in question (say a certain number of tons of coal or so many square feet of undisturbed soil and grass) can be valued in terms of price or opportunity costs. In these cases, the market values (not only to potential buyers and sellers, but occasionally to the public at large) allocate such environmental resources as land use and water rights. (Bureau of National Affairs, 1981b, p. 1114)

The administration acknowledged that it was more difficult for the market to determine the value of clean water and clean air. Even so, the best approach was to "try to use free market principles to assess public attitudes," and the administration used this assessment to find optimal resolutions (Bureau of National Affairs, 1981b, p. 1114).

Security

Similarly, energy security was to be left predominantly to the marketplace. The administration indicated that it would place "primary reliance on market forces to determine the price and allocation of energy supplies even during an energy emergency" (Bureau of National Affairs, 1981b, p. 1113). Reagan spokesmen argued that, left to its own functioning, the marketplace would not only handle the short-term energy security of the nation but develop the needed long-term resources.

In sum, the Reagan administration consistently articulated the view that the nation's energy goals would most effectively be achieved by removing government from the scene. In seeking to implement its philosophy, the administration took a number of actions with regard to the various potential and available sources of energy (see below).

The Congress was not willing to adopt the Reagan philosophy wholesale. The hard-won consensus on energy achieved by 1980 was not easily modified. In two areas in particular the Congress restrained the Reagan initiatives. While agreeing that the energy budget had to share the austerity of other non-defense, nonentitlement areas, Congress modified the administration's proposed expenditure patterns within that budget. Of equal significance, Congress sought to limit the administration's drive to carry out a wholesale dismantling of the energy policy system. In the following pages the differences between the perspectives of the president and those of Congress are noted.

PRESIDENT REAGAN'S ENERGY ACTIONS

The initiatives of the Reagan administration during its first two years in office can be grouped in five categories. In four of those categories the initiatives were clearly defined, though the degree of emphasis given to individual initiatives varied. The four areas were (1) emergency preparedness, (2) deregulation, (3) leasing of federal lands, and (4) federal financial support for energy. In the fifth category, synthetic fuels, the administration's position was less clear.

Emergency Preparedness

The oil shortages of 1973 and 1979 had impelled a search for policies that came to be labeled "emergency preparedness." The overarching goal was to devise effective means of minimizing the adverse consequences of sudden disruptions of oil imports. As discussed earlier, two important steps had been taken toward that goal when the Reagan administration came into power: the Strategic Petroleum Reserve and the standby rationing program. The authority for the latter ended on September 30, 1981, with the expiration of the Emergency Petroleum Allocation Act of 1973.

The new administration supported the SPR. The only reservation was the "budget-busting" costs of purchasing the oil for the reserve. In cooperation with the Congress, the administration quickly moved funding for the SPR into an off-budget category (*Oil and Gas Journal*, 1981). With this financial modification the Department of Energy was able to accelerate the pace at which it was filling the SPR, and by January, 1983, the SPR contained almost 300 million barrels. While this amount represented a marked increase in the reserve that the administration inherited, the rate at which the SPR was being filled still did not satisfy some critics. The rate of 177,000 barrels a day urged by Reagan officials at the end of 1982 appeared much too slow. The critics pressed for a more aggressive policy to reach the national storage goal of 750 million barrels in a shorter time (Mosher, 1983a).

On the other hand, Reagan was absolutely opposed to the second emergency-preparedness measure—the standby rationing program. He believed that even a standby program would send the wrong signal to the market and that in the event of another oil cutoff government efforts to manage a rationing system would be more disruptive than normal marketplace management. Clearly the president and the Congress differed on this point: in 1982 the president vetoed a Congress-approved rationing plan. This bill, the Standby Petroleum Allocation Act, was a mild version of earlier measures. It would have allowed but not required presidential allocation of energy supplies if the crisis was too severe for the market to handle. But even this relatively modest contingency planning mechanism was unacceptable to the administration. Secretary of Energy Edwards argued during the battle to sustain the president's veto, "I am confident that free market forces, unimpeded by federal government interference, will be the best response to any likely energy shortage which might occur" (Bruce, 1983, p. 7). Such a stance, in which reliance is placed on "self-insurance" by the private sector and "volunteerism" by consumers, is regarded in many quarters as the equivalent to an attitude of every man for himself in a crisis. To some experts the administration appeared to be taking a high-risk course.

Deregulation

The Reagan administration sought to eliminate or modify regulations affecting energy in the categories of price, environmental and health protection, and energy performance. Within days of taking office, President Reagan accelerated by eight months the deregulation of oil prices that had been initiated by Carter. The president repeatedly argued that natural-gas prices should be similarly deregulated but delayed for more than two years a formal call for total gas-price deregulation, which finally came in February, 1983.

As the president outlined the plan, price controls on all natural gas would be dropped on January 1, 1986, but in the interim prices would not be allowed to rise as much as under the provisions of the Natural Gas Policy Act of 1978. This temporary restraint on price increases would be accomplished by empowering the Federal Energy Regulatory Commission to limit pass-throughs of higher gas costs to consumers (Williams, 1983). As might be expected, given the history of natural-gas politics, opposition to the proposal was quick in coming. Consumer groups argued that the costs to users would benefit the larger oil companies, which owned much of the "old" gas in the United States (Benjamin, 1983).

Two factors help explain Reagan's two-year delay in submitting the proposal and the difficulties he faced in gaining passage of this or any other legislation deregulating gas prices. First, following the passage of the law of 1978, the nation achieved a substantial natural-gas surplus. Second, opposition to deregulation of gas was broadly based, including many members of Congress as well as some components of the natural-gas industry.

The administration entered office maintaining that excessive environmental regulation was a major reason for the poor performance of the economy and the energy system. The examples most frequently mentioned were the nation's failure to use greater quantities of coal and to build more nuclear power plants. With regard to coal, the administration saw the Clean Air Act Amendments of 1977 and the Surface Mining Control and Reclamation Act of 1977 as major causes of the energy problem. Recognizing the congressional commitment to environmental protection, however, the administration did not submit formal proposals to modify these pieces of legislation. Instead the administration took a different tack. First, environmental and health regulatory agencies experienced large budget and personnel cuts, which were explained as part of the administration's effort to reduce the size of the federal government. Second, the new administrators of the regulatory agencies enforced regulations with less vigor. Third, the administration set out to modify the existing regulations. In the early days of the Reagan administration this strategy led to a moratorium on new regulations. Only regulations specifically approved by the Office of Management and Budget could be imposed. Major proposed regulations were subjected to a federal cost-benefit analysis. Such regulations included those likely to have an annual effect on the economy of $100 million or more, lead to major increases in costs or prices, or have significant adverse effects on competition, employment, investment, productivity, innovation or the ability of American enterprises to compete with foreign-based corporations in domestic or export markets. Decisions to defer, revise, or rescind regulations were made on the basis of this analysis, and agencies had to choose the least costly method of reaching their regulatory goal (see Eads, 1981). This decision-making process began in the Ford and Carter administrations but became much more centralized and comprehensive in the Reagan administration. Few regulations were able to pass the cost-benefit test.

Additionally, all agencies were required to review existing regulations and eliminate those that were unnecessary in the view of the administration and to modify those that were considered too stringent. Perhaps the most striking example was a wholesale revision in the regulations established by the Office of Surface Mining to implement the Surface Mining and Reclamation Act. The stated objective of the revisions was to give the industry greater self-regulatory responsibility and to transfer most of the enforcement responsibility to the states. At the end of two years both the administration and its critics could agree that in surface mining the stringency of environmental regulations and enforcement had been significantly reduced.

The same objectives were pursued with regard to nuclear power, but without discernible success. In his statement on nuclear energy policy of October 8, 1981, the president directed the Department of Energy to recommend ways to streamline the regulation and licensing of nuclear power plants to reduce the licensing time from a lag of 10 to 14 years to 6 to 8 years. Although changes were initiated that provided some flexibility for the nuclear industry, such as permitting the use of interim operating licenses and improving the oversight capability of the Nuclear Regulatory Commission, the administration continued to seek more fundamental modifications in the system (Marshall, 1981).

As noted earlier, with a few exceptions efforts by the federal government to impose energy performance standards were unsuccessful, and by 1980 the Carter administration and the Congress were abandoning the approach. The Reagan administration is distinctive only in its categorical opposition to all government efforts to enforce energy conservation or efficiency. For example, the Reagan administration stated its clear opposition to any extension or tightening of performance standards for cars and called for the termination of performance standards for appliances and buildings and an end to utility conservation services (U.S. Congressional Research Service, 1981).

Thus the Reagan administration sought to weaken regulation of both energy production and energy consumption across the board. For the most part the administration's strategy was to revise and undercut regulation by reducing budgets, eliminating or transferring personnel, rewriting rules, and weakening enforcement. By this means the administration attacked what it considered to be barriers to free-market decision making with regard to energy. At least some of these actions appear to have been designed to bypass expected congressional resistance.

Leasing of Federal Lands

In its early days the Reagan administration expressed the intent to move vigorously to make federal lands available to private-sector energy developers and producers. The administration proposed a significant modification in the leasing procedures of the Department of the Interior on the outer continental shelf oil-and-gas lands. The DOI was to publish a new five-year leasing schedule that would offer the industry a much wider choice of tracts on which it could bid (U.S. Department of the Interior, 1981). The rationale underlying the new schedule was that the industry should determine those areas with the greatest potential for giant new oil-and-gas fields and should be allowed to bid on the areas of their choice.

Previously the DOI had designated specific tracts that would be available for bidding. Under the new arrangements industry could bid on tracts within large planning areas. As part of this new approach the DOI revised the manner in which the department's Minerals Management Service evaluated the adequacy of industry bids. The new approach reduced the role of Minerals Management Service economic evaluations in determining when bids were unacceptably low. Under the five-year schedule approximately one billion acres of OCS lands would be available for bidding.

Efforts were also made to expand leasing on onshore federal lands. The Reagan administration tried to accelerate the process of determining which lands would be designated as wilderness areas under the Federal Land Policy and Management Act. Early determination would allow the rapid leasing of non-wilderness lands. The goal of this program was to increase the amount of land available for leasing and therefore oil-and-gas exploration and production. It should be noted that, in pressing for increased federal leasing, the administration was following a pattern that had already been established.

In another area, however, the administration sought to modify existing patterns. Its intent was to streamline approval for oil-and-gas operations after leasing. Industry had consistently argued that, even when it was able to lease, exploration and development were frequently inhibited by bureaucratic impediments and unrealistic standards.

The program under which federal coal lands were leased

also was significantly modified by the Reagan administration. Under the Carter administration the DOI had sought to determine future needs for coal and offer for lease coal lands sufficient to meet those needs. The Reagan administration rejected that approach and argued that the private sector should determine demand. The administration offered larger quantities of federal coal lands for lease. It also modified the system by which the DOI determined the fair-market value of the coal being mined under lease. The new approach once again placed primary responsibility on competitive bidding to determine the value of the coal and reduced the role of the Bureau of Land Management in determining the adequacy of bids.

In general, the Reagan administration's objectives for public lands were extensions of the policies evolved between 1973 and 1980. What is striking is the intensity of the controversy generated by the administration's management of those policies. The rhetoric of Reagan officials often seemed designed to provoke conflict. When, for example, Secretary of the Interior James Watt said that "conservation is not the blind locking away of huge areas and their resources because of emotional appeals" and blamed "the extreme environmentalists and the immovable machinery of the bureaucracy" for slowing or halting energy resource development on public lands (Congressional Quarterly, Inc., 1982b, p. 6), he was tossing down a political gauntlet. Such statements reduced the potential for consensus-building and increased the likelihood of a return to veto politics.

Federal Financial Support

In no area did the Reagan administration demonstrate its commitment to marketplace management of energy over both the short and the long term quite as clearly as in that of federal financial support for the development of energy resources. Except for the nuclear fast-breeder reactor, where Reagan's initiatives ran counter to the marketplace-as-arbiter notion, and long-term R&D, the administration followed a consistent pattern of proposing massive cuts in federal support for energy.

The extensive changes proposed and implemented by the Reagan forces are clearly illustrated in the R&D budget for energy. Overall, administration proposals for DOE-supported

R&D would have reduced funding approximately 44 percent, from roughly $4.5 billion in FY 1981 to $2.5 billion in FY 1983 (Shapley, Teich, and Weinberg, 1982, p. 10). The budget for energy related environmental R&D funded by the Environmental Protection Agency called for a drop of 65 percent over the two-year period, from about $100 million to $35 million (Shapley, Teich, and Weinberg, 1982, p. 68). Those proposals do not reflect the even greater structural changes in the approach to the financial support for energy. Under the Reagan program three energy sources would have suffered 80 percent of the total cut within the Department of Energy's R&D budget. Those sources were fossil fuels—primarily coal—energy conservation, and solar power. The proposed reduction for fossil-fuels R&D between 1981 and 1983 was from $991.5 million to $106.8 million. For energy conservation R&D the proposed decrease was from $201 million to $19.3 million. For solar energy the suggested reduction was from $549.1 million to $73.5 million (Shapley, Teich, and Weinberg, 1982, p. 145).

In proposing these cuts, the administration argued that the marketplace could more effectively determine the appropriate emphasis that each energy source should receive. Yet that argument was rejected in the proposed R&D budget for the fast-breeder reactor. In his nuclear policy statement the president said that federal subsidies were necessary because excessive federal regulations of nuclear power had severely hampered the industry's ability to accumulate capital and to compete effectively in the marketplace.

It will be recalled that President Carter had sought to terminate the breeder-reactor program but had been blocked by the Congress. The Reagan administration articulated enthusiastic support for the breeder, and its proposed funding for that program called for only a 6 percent drop between 1981 and 1983, with funding at the level of $621.9 million in 1981 and $582.4 million in 1983 (Shapley, Teich, and Weinberg, 1982, p. 38). In addition, the Reagan administration lifted the Carter administration's ban on commercial reprocessing of spent nuclear fuel from commercial reactors (Bureau of National Affairs, 1981c). The reprocessing market was so weak, however, that this action did not appear likely to move the private sector toward developing a commercial reprocessing capacity. By early 1983 the administration had not committed itself to providing any government support for spent-fuel reprocessing, though it had taken under consideration a recommendation that it provide an "improved regulatory and licensing arrangement," purchase guarantees for plutonium, and investment supports to ensure completion of the Barnwell, South Carolina, facility (Benjamin, 1982). Contrary to its position on the breeder reactor, the Reagan R&D proposals would have reduced federal support for conventional nuclear-reactor development from $84.3 million in 1981 to $33.6 million in 1983.

In only one energy policy area, long-term energy research, did the Reagan administration propose increased federal support. Funding for basic research, primarily in high-energy physics, would have grown by over 15 percent between 1981 and 1983, from a level of $521.8 million to $601 million (Shapley, Teich, and Weinberg, 1982, p. 38).

The Congress was unwilling to accept many of these proposals, particularly those for fossil fuels, conservation, and solar power. The differences in the proposed budget figures and those actually approved by Congress reflected basic differences in philosophy. Some members of Congress continued to see energy as too important to be left to the whims of an imperfect marketplace. Thus in the FY 1983 budget the support levels provided by the Congress were as follows: fossil fuels, $289.2 million; conservation (including both R&D and subsidies), $343.3 million; and solar energy, $276 million (U.S. Congress, House of Representatives, 1982). Congress thus appropriated more than the administration's final requests by the following amounts: fossil fuels, $182.3 million; conservation, $321.5 million; and solar energy, $203.8 million.

The energy-budget disputes between the administration and the Congress clearly represented not only philosophical but also strategic differences. Traditionally budget struggles have involved congressional fine-tuning of presidential proposals, the major differences centering on individual projects. Between 1981 and 1983, however, the struggles were over whether or not to eliminate the fossil-fuels, conservation, and solar programs. Congress acted to save them.

By early 1983 the battle appeared far from over. In its proposals for FY 1984 the Reagan administration did not veer from its path. In its analysis of the R&D budget provisions for FY 1984 the Office of Management and Budget said:

Obligations for the conduct of R&D programs [including nuclear weapons R&D] currently in the Department of Energy are estimated at $4.7 billion, about the same as in 1983. However, basic research funding would be increased by $160 million, or more than 18 percent over 1983. Increases are proposed to strengthen the nuclear weapons R&D program, to enhance support for long-term energy research, and to maintain a strong national basic science effort in high energy and nuclear physics. These increases are offset by proposed reductions in support for nearer-term energy technologies such as fossil fuel and solar demonstrations, where reliance is placed on private sector support. (U.S. Office of Management and Budget, 1983, p. K-4)

In fact, one project, the breeder reactor, would have consumed about $603 million if the 1984 budget proposals were accepted. This amount, an increase of about $62 million from 1983, would have exceeded the entire proposed expenditure for fossil fuels, solar energy, conservation, and biological and environmental research. In the 1984 budget initiative fossil, solar, and conservation programs were to be funded at $302 million, a decrease of $405 million from 1983. Fossil-fuels R&D alone were to absorb a cutback of $234 million in this proposal (U.S. Office of Management and Budget, 1983, p. K-14).

Synthetic Fuels

Only one area of the Reagan administration's approach to federal financial support for energy remained unclear. This was the area of synfuels development. Early indications were that the administration was opposed to the very existence of the Synthetic Fuels Corporation, and that view was reinforced when Edward Noble, an Oklahoma oilman, was appointed chairman of the board of the SFC. Noble indicated publicly that he had not supported the establishment of the SFC in the first place. After becoming chairman, however, Noble reversed his view (Hennrich, 1982). He expressed the belief that the corporation should underwrite the development of synthetic-fuels technologies because someday synfuels would be needed.

273

President Reagan's first action in this area was to approve federal support for three large synfuels projects, two oil-shale ventures in Colorado and a project to generate synthetic gas from coal in North Dakota. These projects had been started by the Department of Energy during the previous administration, and an immediate decision on their future was necessary. The president apparently was convinced that support for the projects should be continued to meet an implicit obligation to the development companies. Although the chairman of the SFC indicated a need for an active government role in synfuels development, after two years the administration had undertaken no new synfuels projects. Clearly the corporation had given up the effort to meet the daily production goals set in the synfuels legislation of 500,000 barrels of oil-equivalent by 1987 and 2 million barrels of oil-equivalent by 1992. Also the administration had indicated that it did not intend to request the additional $68 billion that could be appropriated under the original legislation (Corrigan, 1982b).

Thus the administration's rhetoric about the SFC was unclear, and its failure to act indicated at best a go-slow attitude. The conclusion of the authors of a Congressional Research Service report was that "no consistent program regarding synthetic fuels development has emerged from the Reagan Administration" (Congressional Research Service, 1981, p. 33).

Impacts on the Energy Policy System

According to Craufurd Goodwin (1981, p. 97), throughout American history energy policy has been "continuously torn between two extremes of economic theory: free markets and central planning." The energy policy system that had been established by 1980 was based on a fragile, sometimes crazy-quilt balancing of these two very different sets of theory. The system had no neat ideological underpinnings; it was based in large part on pork-barrel politics and trade-offs. Nevertheless, a minimal consensus had been achieved. In its pursuit of marketplace management the Reagan administration undercut this consensus.

The Energy Policy Community

A brief review of what happened to the energy policy community after 1980 demonstrates the administration's success. The numbers of participants from the federal government, state governments, the private sector, and nonprofit research organizations were greatly reduced. The new administration's prime intent was to constrict the role of federal agencies in the energy system. For Reagan the Department of Energy symbolized unwarranted intrusion of the federal government into the private sector. During his election campaign candidate Reagan had committed himself to the elimination of the DOE, and his first secretary of energy, James Edwards, went to Washington determined, in his words, to "close the Energy Department down and work myself out of a job" (Congressional Quarterly, Inc., 1982a, p. 4). This commitment was translated into a formal legislative proposal on May 24, 1982, when the administration asked for dissolution of the department and the transfer of many of its functions to the Department of Commerce, claiming that this structural "reform" would save about $1.3 billion in FY 1983. As the proposal was more closely reviewed, however, the administration lowered its assessment of savings to $250 million. Even that figure was said to be too high in a General Accounting Office report issued in August, 1982 (Bureau of National Affairs, 1982a).

By the end of 1982, although the Congress refused to dissolve the DOE, it experienced significant reductions in personnel and funding. The department's energy R&D budget was cut nearly in half. Moreover, during the tenure of Secretary Edwards the DOE lost about 3,500 positions, or about 17 percent of its work force. Most of the department's regulatory activities were either significantly reduced or eliminated, which led one observer to note that Edwards "was not asked to do very much at the Department except to immobilize it if he could, and he carried out his assignment in that spirit, as a negative force at the top favoring institutional non-action and self-destruction" (Corrigan, 1982c).

Most analysts believe that there was little more in the way of financial savings to be achieved from the elimination of the DOE because its remaining functions would have to be transferred to other agencies if it was eliminated. In fact, the administration's progressive weakening of the department became such a concern to the Congress that it took the highly unusual step of barring further reductions in staff in a supplemental appropriation to the FY 1982 budget (Appropriations for DOI and Related Agencies Act, 1981). By that time, however, the DOE's role in energy policy had been reduced to a peripheral one, leaving the debate about its elimination little more than symbolic.

The administration dealt with the Environmental Protection Agency in a similar fashion. As previously noted, the EPA's proposed energy-related R&D budget for 1983 would have been cut by 65 percent, and the administration proposed to reduce the staff by about 25 percent (Mosher, 1982a). Not surprisingly, these proposals raised heated questions about the administration's intentions on enforcing environmental-protection laws. In the eyes of administration critics, "EPA is moving in exactly the wrong direction now. . . . We simply cannot tolerate the size of these cuts" (Mosher, 1982b, p. 637). Congress viewed the situation as so serious that it passed legislation to increase the EPA's research authorization and to direct more spending toward long-range projects. The president vetoed the measure (Mosher, 1982c).

In the vacuum created by the weakening of the DOE and the EPA and under the aggressive leadership of James Watt, the Department of the Interior took on a more important role in the energy policy system. As we have seen, however, that role was largely confined to accelerating the leasing of federal lands. The environmentally oriented branches of the DOI experienced treatment similar to that of the DOE and the EPA, particularly the Office of Surface Mining and its regulatory responsibilities. Some members of Congress viewed the Reagan administration's efforts to reshape OSM with so much concern that the House added provisions to the Interior appropriations bill of 1982 that forbade the reorganization of the OSM without explicit congressional permission (Appropriations for DOI and Related Agencies Act, 1981).

The DOI sought to emphasize energy resource development on the OCS and on federal lands through two further reorganization actions. It created the Minerals Management Service to streamline OCS development and moved the pre- and post-lease components of the U.S. Geological Survey, which had previously managed onshore energy development on federal lands, to the Bureau of Land Management. Interior Secretary Watt explained that this reorganization would ensure that the BLM would have a permanent prominerals-development component.

Of the energy-related regulatory agencies only the Nuclear Regulatory Commission was maintained in its pre-Reagan form. The administration's position has been that working out an ordered, careful program is necessary to assure larger amounts of electricity from conventional nuclear power.

Finally, the Synthetic Fuels Corporation continued operations under the new administration, but at a much slower pace than was called for in the legislation or expected under the Carter administration. The new president, of course, replaced Carter appointees in the corporation with his own people (Plattner, 1982, p. 1250).

Participation by the states in the energy policy community developed primarily from incentives provided by federal programs. In the 1970s many state agencies were created to enforce federal environmental legislation and to implement federal energy-conservation programs. Most of the funding for those agencies came from federal sources. The Reagan administration began trying to pass more responsibility to the states, meanwhile cutting back federal support for energy conservation and environmental-protection activities at the state level. For example, the EPA had provided assistance to the states through grants-in-aid and other forms of financial, technical, and information support. The Reagan administration reduced this support steadily between 1981 and 1983. Total grant support declined by about 18 percent (it would have declined 34 percent if Congress had accepted all of Reagan's proposals). Moreover, Ann Gorsuch (later Burford) indicated that the "EPA intends over the long term to phase out state grants entirely" (*Inside EPA*, 1982). Both the National Governors' Association and the National Conference of State Legislatures maintained that significant reductions would force states to reduce environmental-protection efforts or even abandon them (National Governors' Association and National Conference of State Legislatures, 1982, pp. 92–94). The overall effect on energy of Reagan's "new federalism" appears to be a reduced capacity by state governments to participate in energy policymaking and implementation.

In a similar pattern, before Reagan's election the Department of the Interior had sought to develop procedures that would allow for increased state involvement in federal leasing, especially of coal lands. The new administration tried to downplay such participation. The effect of these policies was a marked decrease in the capacity of state governments to take part in national energy policy and administration.

In the private sector the new administration showed a marked preference for energy-producing companies, seeking to exclude environmental interest groups from the decision-making process. Unlike the past three administrations, the Reagan administration felt no obligation to seek input from environmentalists on decisions about energy-resource development. Environmental groups generally perceived themselves as having very little influence on national energy decision making under the new rules of the game (Friends of the Earth, 1982). Finally, the substantial reduction in federally sponsored energy R&D lessened the role of private-sector research groups in energy policymaking. Nonprofit R&D programs, particularly those at universities, also felt the effects as support was withdrawn from their researchers.

SUMMARY

A review of the actions proposed and taken under Reagan administration leads to one clear conclusion: the administration saw no need for a discrete national energy policy or an energy policy community. Primary responsibility for energy policy was handed over to the private sector, with policy management to be handled through marketplace decisions.

With regard to national energy goals, the Reagan administration also committed itself to the marketplace process. It downgraded emphasis on environmental protection and therefore the clean-energy goal. The administration's approach to the goal of energy abundance was to seek to make publicly owned resource lands more rapidly available for development by the private sector. Finally, the administration articulated the view that the goal of energy security was most effectively achieved by allowing the free working of the marketplace.

Thus the Reagan administration dismantled the energy policy community that had come together between 1973 and 1980. Within months the role of the federal government had been reduced to two areas, the leasing of federal lands and support for the breeder reactor. Its approach to synfuels development was unclear.

Federal energy policy would have been even more seriously eroded had it not been for congressional resistance. Over administration objections Congress retained federal programs in fossil fuels, conservation, and solar energy. Congress also resisted modification of environmental laws and sought to provide some protection for federal energy-related agencies.

PART FOUR

Energy Policy for the Future: Conclusions and Recommendations

CHAPTER 12

The Need: A Return to Traditional Policymaking

The energy argument has become an end in itself, but what is needed is an honest consideration of all the options if we are to have fuel instead of ideology. —John O'Brien, *Bulletin of the Atomic Scientists*, April, 1982

The initial statement about the purpose for this book characterized it as an investigation of the nation's failure to formulate an effective energy policy following the oil embargo. As this chapter is written, it remains appropriate to characterize the search for an energy policy as a failure. The nature of that failure, however, is much different from what was initially perceived. Our most striking finding while researching this book was that the nation's policy processes had established a workable energy system by the end of 1980.

In the Reagan years the nation has witnessed failure snatched from the jaws of success. The source of the failure is the Reagan administration's rejection of the traditional processes of American policy formulation, implementation, and management.

As noted earlier, in the traditional American policymaking process the president and the Congress formulate policy and establish semiautonomous systems to implement that policy. As long as policy systems function successfully, that is, maintain stability, presidential-congressional involvement is limited to ratifying policy changes around which a consensus has evolved. Only when instability is introduced because something goes wrong with the way nationally important physical activities are managed are those activities returned to the presidential-congressional agenda, where the policy system is redefined or a new one is constructed.

What occurred in the field of energy in the 1970s was a particularly important instance of the workings of the nation's traditional policy processes. The oil embargo produced policy instability by mobilizing demands for participation in energy policymaking by many actors and interests, and it spotlighted fundamental issues that came to be labeled the "energy crisis."

In the early 1970s the energy crisis was an event waiting to happen. To recapitulate: In the years immediately preceding the embargo the changing facts of energy had rendered existing fuel policies and policymaking systems obsolete. The organization of policy around specific fuels made it impossible for the nation to anticipate and effectively respond to scarcity of domestic energy. When some of the nation's oil imports were cut off, it became obvious that the nation did not have the capacity to arrange trade-offs among conventional fuels, manage demand, or develop new sources of supply.

In a manner consistent with traditional policymaking processes, the search began at the presidential-congressional level for a policy and a policymaking system capable of managing the new facts of energy. What was needed was a system that could integrate and manage twelve existing and potential energy resources and a large number of actors and interests, some of which had major equity and environmental concerns. Long-term success would be achieved through a problem-solving, consensus decision-making process.

By the end of 1980 the essential components of a stable energy policy system were in place. The goal of cheap energy had been rejected, and the goals of abundance, cleanness, and security had been redefined so that they were consistent with the facts of energy. Decisions had been made to move conservation and synfuels into the category of available resources. A program of federal R&D support for potential and theoretically possible energy resources was in operation. The decision had been made to allow OPEC oil prices to set the ceiling for domestic-energy prices. In addition, the nation had chosen to use federal subsidies as energy incentives, rejecting the use of sanctions. Finally, the main continuing participants in the energy policy system had been sorted out. A rudimentary policy system had been constructed to manage an energy sector that included twelve possible energy resources. Thus in the course of less than a decade the nation had moved from stable energy management through five fuel policy systems to a period of instability during which energy was the province of the president and Congress to at least the framework for a stable energy policy system.

At this point the Reagan administration launched the country on a radical new experiment. The new administration saw no need for a national energy policy or for an energy policy system. With limited exceptions the administration also saw no specific role for the federal government or "outside" actors or interests in energy. The administration was convinced that energy should be managed in the marketplace, through the president's "Program for Economic Recovery."

The administration's commitment to marketplace management as the way to achieve abundant, clean, secure energy resulted in a massive withdrawal of federal support for the development of new sources of energy for both the short and the long term. The free-market approach resulted in a dismantling of the energy policy community. Only the producers of the conventional energy resources remained; the earlier participants were either reduced or eliminated, not only state and federal agencies but also environmental interest groups, R&D companies in the private sector, and nonprofit research institutions. The prospect of stable energy policy and management had vanished.

Both its critics and its defenders agree that the Reagan administration's approach to the management of energy was a decisive and radical break with the past. Not everyone, however, agrees with our conclusion: that reliance on the marketplace has put this nation at risk, with the prospect of a highly unstable energy future. In the following pages we offer our reasons for this forecast and make some suggestions for alternatives.

As discussed in chapter 2, the record of success of semiautonomous physical policy systems goes back to the origins of the nation. In truth, the nation's history of political stability derives from the stability provided by the collection of semiautonomous policy systems that have made and managed physical activities. That record of stability provides powerful evidence that the nation should not abandon its traditional policy processes.

What, then, could be the rationale for a switch to marketplace management? Was there practical (or historical) evidence of successful marketplace management of energy activities? Was there a theoretically compelling basis for believing that the marketplace could best serve the nation's energy goals?

Chapter 3 reviewed the history of public policy for the five fuels that were providing the nation's energy at the time of the embargo. It will be recalled that the national goals served by those policies were cheap energy and abundant energy. An overwhelming proportion of the five energy fuels was produced and supplied by private-sector profit-making companies. The record of success of the fuel policy systems in delivering cheap, abundant energy is evidence that the private-sector producers enjoyed adequate profits. It is important to note that physically oriented policy systems by no means preclude attractive profits for the private sector. Quite the contrary, one of the purposes of those systems traditionally has been to assure profits.

The record of all the fuels except coal was one of success and stability until shortly before the oil embargo. Repeatedly, when the stability of those systems was threatened, policy was modified to reestablish stability and assure the continuing availability of cheap, abundant supplies of energy with satisfactory profits to the producers.

In oil, public policy was used to put a floor under domestic prices by controlling production. When the need for imports grew because of insufficient domestic production, the policy system was adjusted to ensure adequate supplies of imports while providing profit protection for the domestic industry.

In the monopolies of electricity and natural gas, public policy was used to assure consumers cheap energy while providing assured profits to the gas and electric utilities. Nuclear power was entirely a creation of public policy through the use of tax monies and federal regulation.

Only in coal did the nation have something of a free market—and the experience of coal does not make a compelling argument for the advantages of marketplace management. By general agreement coal was the "sick man" of the nation's energy system.

Despite the claims of advocates of marketplace decisions that all we have to do is "get government off our backs" and Btu's will immediately be forthcoming, there is no evidence that such magic has ever taken place in the energy system. Daniel Bell's analysis comes closer to the actual experience:

There is economic logic to the release of price controls—if phased out so that consumer and industrial markets are not disrupted. And this is being done. The price of a product should reflect its full economic—and in the case of oil, its full political—cost, otherwise there are serious disruptions in the pattern of use. But that is a different matter from saying that the release of prices will easily stimulate large new supplies. . . . The evidence we have seen does not warrant such a claim. (Bell, 1981)

In sum, there is no instance in which the marketplace has effectively managed energy even when public-policy goals were limited to cheap and abundant energy.

With the addition of the goals of clean and secure energy, even orthodox economic theory militates against reliance on the marketplace. Economists label clean and secure energy as a public good. By general agreement market prices and costs, the essential figures used by private investors to evaluate return on investment, do not take these broader social values into account. National security and environmental protection are prime examples of values that are not normally reflected in private-sector, marketplace decisions. Private-sector decision makers enjoy no benefits from calculating the costs and benefits of national security and environmental protection. Only broader policy systems can integrate security and environmental protection in decision making.

Assuring abundant sources of clean, secure energy requires long-term policy management that provides a stable, predictable energy environment. Private-sector calculations of profit must inevitably be responsive to short-term economic forces, particularly in recessionary periods, as opposed to long-term social, economic, and political goals. The time constants of corporate decision making are, at best, a few years. Planning and decision making in the private sector must be responsive to the annual (frequently quarterly) needs for profit. Anything beyond five years falls in the category of long-range planning and is susceptible to continuous change. Thus both the time horizons and the values that predominate in the marketplace mean that, if long-term goals such as national security and environmental protection are addressed at all, it is by chance rather than by conscious choice.

Since 1973 the United States energy market has operated in

a setting in which both supply and price are pawns of changing political and economic objectives or alternatively political instability in and among producing countries. Hovering over United States energy decision making and management is the OPEC phenomenon. The oil embargo was imposed by the Arab members of OPEC as a self-conscious effort to use oil as a political instrument. The oil disruption resulting from the revolution in Iran was the result of political struggles within that country. In a world where both the price and the supply of oil are to a greater or lesser extent the product of changing policy or instability in exporting nations, an approach that rests on the ideal that the marketplace can provide for the nation's needs seems indefensible.

To summarize: In a time when the United States remains heavily dependent on imported oil and there is no free world market in oil, faith in marketplace management of energy seems foolhardy. Moreover, it does not take into account such public goods as national security and environmental protection or the short-term decision making characteristic of the private sector. The idea that the nation will place primary reliance on an abstract conception of a free market to manage its lifeblood carries unacceptably high risks. Yet that is precisely the approach taken by the Reagan administration. Reliance on the market denies the nation a conscious public policy aimed at developing domestic alternatives to oil, and, perhaps more important over the long term, it precludes the development of an energy policy system capable of providing stable management given the changed facts of energy.

THE SITUATION IN 1983

What did the facts of energy look like in 1983? Certainly they were complex and laced with uncertainty. Chapters 5 to 7 presented a summary of those facts. Only established technologies used to produce energy from conventional sources (oil, gas, coal, electricity, and nuclear power) were available and economically viable. In 1983 with minor exceptions time and cost availability for all other energy resources remained as they had been in 1973, subjects of controversy and uncertainty within the technical community. The future availability of oil, gas, and uranium was in doubt because of the difficulty in determining the size of the resource base. There was also uncertainty about the contribution that energy conservation would make in the future and therefore about future energy needs.

Thus the major change in the facts of energy was that uncertainty had replaced the confidence of the years up to 1973. The most compelling reason for establishing a stable overall energy policy system was the need to manage uncertainty.

Although uncertainty about the facts of energy remained high, in a number of areas the picture was more optimistic in 1983 than it had been in any year since the embargo. Following is a brief summary of what was generally perceived to be the energy situation in early 1983.

First and perhaps most striking was the energy contribution obtained from the increasingly efficient use of energy—from conservation. After 1973 the nation developed the capacity to enjoy economic growth with ever-smaller quantities of energy (Editorial Research Reports, pp. 13–14). At the time of the embargo it took approximately 59,000 Btu's of energy to produce a dollar's worth of economic activity. In 1983 that figure was estimated to have declined to 48,000 Btu's (Beck, 1983, p. 78). By one calculation, if every unit of gross national product generated in the United States had required the same amount of energy in 1981 as in 1973, the country would have consumed the equivalent of 7.3 million barrels a day of oil, more than it actually consumed in 1981 (Yergin, p. 125). Further, it was widely assumed that energy conservation would continue to make a contribution to the nation's energy needs. That is, energy conservation was not viewed as a short-term phenomenon. There were likely to be continuing, incremental energy savings.

The change in predictions of energy consumption owing to conservation was strikingly presented by Amory Lovins (see table 12.1). Lovins pointed out that, no matter what the bias of the forecaster, all energy-demand predictions for the year 2000 had been dropping at about the same rate since the oil embargo. He divided the forecasts into four groups: "beyond the pale" (made by low-growth advocates like himself), "heresy" (put out by moderate conservationists), "conventional wisdom" (including government scenarios), and "superstition" (made by high-growth advocates). Lovins noted that what was beyond the pale in 1972—an expected 125 quadrillion Btu's of demand at the end of the century—became mere heresy in 1974, conventional wisdom in 1976, and superstition in 1978 (quoted in Marshall, 1980, p. 1354).

Table 12.1. U.S. Energy-Demand Forecasts, 1972–78

Year of forecast	Beyond the pale	Hersey	Conventional wisdom	Superstition
1972	125 (Lovins)	140 (Sierra)	160 (AEC)	190 (FPC)
1974	100 (Ford zeg)	124 (Ford tf)	140 (ERDA)	160 (EEI)
1976	75 (Lovins)	89–95 (Von Hippel)	124 (ERDA)	140 (EEI)
1977–78	33 (Steinhart)	67–77 (NAS I, II)	96–101 (NAS III, AW)	124 (Lapp)

Abbreviations: Sierra, Sierra Club; AEC, Atomic Energy Commission; FPC, Federal Power Commission; Ford zeg, Ford Foundation zero energy growth scenario; Ford tf, Ford Foundation technical fix scenario; Von Hippel, Frank Von Hippel and Robert Williams, of the Princeton Center for Environmental Studies; ERDA, the Energy Research and Development Administration; EEI, Edison Electric Institute; Steinhart, 2050 forecast by John Steinhart, of the University of Wisconsin; NAS I, II, and III, the spread of the National Academy of Sciences Committee on Nuclear and Alternative Energy Systems (CONAES); AW, Alvin Weinberg, study done at the Institute for Energy Analysis, Oak Ridge; Lapp, energy consultant Ralph Lapp.

Source: Eliot Marshall (quoting Amory Lovins), "Energy Forecasts: Sinking to New Lows," *Science* 208 (June 20, 1980):1353, 1354, 1356. Copyright 1980 by the American Association for the Advancement of Science.

Second, the gas shortages of 1976 and 1977 had disappeared. With the passage of the Natural Gas Policy Act, which allowed the price of gas to increase and eliminated the distinctions between intrastate and interstate markets, a gas surplus developed. The increasing price of gas also triggered substantial exploration, and the belief was widely held that new discoveries of gas would be more than adequate to take care of the nation's needs for the next two decades. Some optimists expected that the discovery of very deep gas and the development of unconventional gas might in fact mean that gas could take over a substantially larger share of the nation's energy market.

Third, coal production, while not increasing as rapidly as proposed by several presidents and as desired by the National Coal Association, had nonetheless increased significantly. Coal production in 1982 was more than 40 percent above the level of 1973. Further, coal was supplying a larger percentage of the nation's total energy needs, having grown from 18 percent in 1973 to 22 percent in 1982.

The annual growth in demand for electric power declined from a rate of 7 percent per year before 1973 to about half that rate by 1982. In that year the electric power industry nationwide, as a result of declining growth in demand, had nearly twice the surplus peak generating capacity that it had traditionally tried to maintain. The industry rule of thumb for surplus peak generating capacity had been 18 percent, and that surplus capacity was at approximately 32 percent in 1982. The electric power industry appeared capable of meeting the nation's electricity needs without great difficulty for at least the short-term future.

Finally, nuclear power was expected to make substantial contributions to the nation's electric power needs as already committed conventional light-water reactors came on line. Although no new conventional reactors were being ordered and there was uncertainty about second-generation breeder reactors, the uncertainty over the future of nuclear power as a whole was much less a concern in 1983 than it had been earlier. Declining demand for energy and the availability of coal made the need for nuclear power appear less compelling. Thus the sense of urgency surrounding nuclear energy decisions was much reduced.

In sum, there was optimism in 1983 about conservation, gas, coal, and electricity. It must be emphasized, however, that expectations continued to be laced with a substantial amount of uncertainty. Everything from how much energy conservation would deliver in the future to what the demand for electricity would be was speculative.

The picture with regard to oil was quite different, however. Serious, pervasive uncertainty continued about the future of United States oil supplies. Estimates suggested that the United States would remain heavily dependent on petroleum imports for the foreseeable future unless domestic liquid alternatives were developed. Projections of continued heavy dependence on imported oil were based on the expectation that domestic petroleum production would experience steady, continued decline. That decline would likely parallel the rate at which the nation was able to reduce its petroleum consumption through conservation, leaving the nation very dependent on imports.

The congressional Office of Technology Assessment (OTA) estimated that conservation and switching from oil to other fuels in the stationary-use area could save up to 1.8 million barrels of oil a day by 1990, with the possibility of significant additional savings after 1990. These savings, plus a possible savings of 0.6 to 1.3 million barrels a day through more efficient transportation, were real reasons for optimism (U.S. Office of Technology Assessment, 1982, p. 185). Unfortunately, those numbers roughly paralleled estimates of the decline in domestic production.

A fundamental and seemingly intractable problem was the inability of the transportation sector to switch to nonliquid fuels. Transportation uses slightly more than one-half of the nation's petroleum. No practicable alternatives to liquid-fueled vehicles were anticipated in the foreseeable future. Similarly, about one-fourth of the nation's petroleum consumption goes into the production of oil-based products. This highly valued use of petroleum is unlikely to shrink by much.

A brief review of what occurred during the 1970s suggests what can be expected over the next few decades. In 1972, the last full year before the oil embargo, the United States consumed an average of 16 million barrels of petroleum a day. Of that amount 11.2 million barrels were produced domestically, and 4.7 million barrels were imported. The nation faced its worst import situation in 1978, five years after the embargo. In that year petroleum consumption in the United States averaged 18.6 million barrels a day, and imports rose to 8.2 million barrels a day (Stobaugh, p. 18). In 1982, in a period of serious economic recession, United States consumption declined to 15.3 million barrels a day. Domestically produced petroleum supplied 10.2 million barrels, and imports 5.1 million barrels (Beck, 1983, p. 75). After 1978, then, petroleum consumption in the United States experienced a steady downward trend, the figure for 1982 being 19 percent lower than the figure for 1978.

It should be emphasized, however, that even the striking reduction in petroleum demand owing to conservation and a recession did not significantly reduce its heavy dependence on imported oil. Estimates of what was likely to happen in the future were at best educated guesses. Most of those estimates, however, projected that imports would remain at about the level of 1982. In a report in that year the OTA summarized its findings as follows:

Plausible projections of domestic oil production—expected by OTA to drop from 10.2 million barrels per day in 1980 to 7 million barrels per day by the year 2000—suggest that oil imports could still be as high as 4 to 5 million barrels per day or more by 2000 unless imports are reduced by a stagnant U.S. economy or by a resumption of rapidly rising oil prices. Achieving low levels of imports—to perhaps less than 2 million barrels per day within 20 to 25 years—is likely to require a degree of success . . . that is greater than can be expected as a result of current policies. (U.S. Office of Technology Assessment, 1982, p. 5)

Two developments could significantly modify that projection. As the OTA noted, any sudden, large increase in the price of petroleum would likely produce substantial increases in conservation. Certainly a repeat of the price increases experienced at the time of the oil embargo and the revolution in Iran would likely reduce demand significantly. Unfortunately, associated with that reduced demand would be substantial disruption of both the national and world economy. Most observers believed that the two previous substantial price increases contributed significantly to worldwide inflation and economic recession. The forces likely to cause substantial reductions in demand carry with them costs that society would surely like to escape.

The second alternative would be the discovery of significant new domestic oil fields. It is important to note that, even though the deregulation of oil touched off a rapid increase in exploration, other than the Western Overthrust Belt and a new field on the California OCS, no new major oil-production areas had been found by 1983. To turn the national production picture around, discovery of more Prudhoe Bays would be necessary. That field was supplying more than 1.5 million barrels a day in 1982. Without Prudhoe Bay, domestic production would have been approximately 8.7 million barrels a day, and imports would have been much higher.

The clearest way to characterize the nation's domestic-pro-

293

duction situation is to note that, at a production level of 10.2 million barrels per day, 3.7 billion barrels of domestically produced petroleum are consumed each year. Simply to sustain production at that level, the United States must discover 3.7 billion barrels of oil each year, or the equivalent of a new Prudhoe Bay field every three years. Few observers expected that kind of discovery rate.

Given the limits on petroleum conservation and the essentially pessimistic expectations about new domestic discoveries of petroleum, the key question was, Did marketplace management of energy provide the nation with adequate security? In the short term, energy security depended on the nation's ability to deal with a denial of imported oil or a sudden increase in the price of imports. As discussed in chapter 10, by the end of 1980 security had come to rest on answers to three questions. First, did imports threaten the nation's ability to protect itself in the event of a military conflict? Second, did the nation's dependence on imports make it susceptible to the use of those imports as a political instrument to press for changes in United States foreign policy? And finally, would the nation's dependence on imports pose a threat to the economic stability of the United States if sudden excessive changes in price or disruptions in supply should occur? Unless it was possible to answer all of these questions with a confident no—and that was not possible—the Reagan administration's approach to energy had to be seriously questioned.

One illustration of the unstable situation with regard to oil can be seen in what happened to liquid stocks (private storage and the Strategic Petroleum Reserve) during 1982. These stocks represented the nation's emergency reserve to meet a sudden disruption in oil supplies. In combination the nation's stocks declined by an absolute 119 million barrels between the end of 1981 and the end of 1982. While SPR stocks rose by 63 million barrels, private stocks declined by 182 million barrels (Beck, 1982, pp. 184, 189; Beck, 1983, p. 73).

In addition, as the situation stood in early 1983, the United States had no domestic liquid-fuel alternatives to imported oil over the midterm, and none was seen on the horizon.

The lack of available domestic liquid-fuel options and a policy to develop those options placed the nation in a high-risk situation in both the short and the midterm. Another sudden disruption of supply and escalation in price of imports could occur at any time. Lacking potential alternatives, the nation had no plan to counter such an event other than the Defense Department's Rapid Deployment Force. Moreover, in the longer term, oil is a finite resource. Alternatives to oil had to be found. On that point there was little disagreement. The search for those alternatives is imperative and should be a self-conscious focus of public policy.

The situation with regard to oil, then, was not significantly different in 1983 from that in 1973. The United States was little better prepared to respond to a cutoff of oil or a rapid increase of price than it had been at the time of the embargo. Thus the posture of the United States in 1983 with regard to oil must be characterized as a policy failure.

CHAPTER 13

Recommendations

Given the grim picture of the nation's energy future presented in the last chapter, what needs to be done? Is there any basis for believing that the United States can in a reasonable period of time reduce its condition of high risk? We believe that there is a real basis for optimism about the future. That optimism is derived from the refusal of Congress to modify the energy legislation passed between 1973 and 1980. Actions by the Congress with regard to energy in the years after 1980 offered compelling evidence that its basic conception of how energy should be managed was unchanged in fundamentals. The necessary ingredient for reestablishing an energy policy system was either a change in approach by the Reagan administration or a replacement of that administration. The relative speed with which it was possible for the administration to erode, disrupt, and dismantle energy policy in the course of two years clearly demonstrated that before an energy policy could work there had to be presidential support for it.

The situation described in chapter 12 indicates the urgent need for a change. When that change occurs, as it must, an agenda of needs for a stable energy system must also be clearly laid out. What follows is a proposal for that agenda.

ENERGY POLICY PARTICIPANTS

Central to reestablishing stable energy management is the reintegration into the energy policy community of all the actors and interests who have demonstrated a continuing commitment to participation in energy policy making. Perhaps the greatest cost in the radical changes after 1980 was the alienation of the environmentalists. Reconstructing an energy policy system will require that the president explicitly and self-consciously support and pursue the clean-energy goal.

Arrangements must be made for assuring continuing participation by environmentalists. At the organizational level the Environmental Protection Agency and the various state environmental agencies must be reinvigorated. At a more general level the president must make certain that managers in the energy policy system, wherever they may be located, again make it a standard order of business to integrate environmental interests into the policymaking process.

Although the Reagan administration has generally sought to exclude the environmental movement from energy decisions, that movement has sustained itself since 1980. In their promotion of a clean environment the environmental organizations reflect a goal broadly supported by Americans. Both to ensure protection for the nation's environment and to ensure that the environmentalists are not to be an external force creating continuing long-term instability, they must be made an integral part of the nation's energy policy system.

An equally important need is to remobilize the scientific-technical talent that was involved in energy research and development before 1981. In the following pages specific energy R&D recommendations are made. The point to be emphasized here is that most energy R&D will be paid for with federal funds. A fruitful, stable, productive R&D system requires the establishment of federal executive agencies that can provide ongoing stable support for it. Stable support is necessary to assure that energy research and development will attract and maintain the high-quality talent the nation needs.

Both the skeleton of the organizational system and the legislative base for stable energy policymaking and management remain. The critical disruptive variable has been the Reagan administration's budgetary and administrative actions. There is substantial reason for optimism that long-term stable energy management can be made available to the nation within a reasonable period of time.

In the following pages recommendations for specific policy actions to reestablish a dependable energy future for the nation. These actions are divided into four categories: (1) short-term, (2) liquid-fuel, (3) legislative-regulatory-managerial, and (4) energy R&D actions.

SHORT-TERM ACTIONS

Recommendation 1: *The Strategic Petroleum Reserve should* *be filled at a rate equal to or greater than that specified in the Emergency Energy Preparedness Act of 1982; that is, no less than 300,000 barrels a day should be added until the combined public and private stocks in the United States are 2 billion barrels or the equivalent of one year's imports.*

Recommendation 2: *In the case of a crisis SPR and private oil stocks should be handled as a single national reserve. A publicly announced industry-government crisis allocation program should be immediately formulated to ensure that the reserve is used as a balance wheel in the case of a crisis. Responsibility for implementing the allocation program should rest with the oil industry, since only the industry has the technical and managerial capability to assure stability in the oil-supply system.*

Instability is an ever-present threat to the world oil market. Friction and conflict characterize the political situation within and among many of the major oil-exporting nations. Prudence demands that the United States prepare itself for the possibility of a third and conceivably a fourth and fifth disruption of oil imports. The conditions that led to the Arab oil embargo and the revolution in Iran have changed in no fundamental way. An oil disruption could occur tomorrow, next month, next year, or five years hence. It is important to emphasize that global political instability is at the root of oil cutoffs. The nation cannot afford to be sanguine about the availability of imports, even in periods of market surplus. The United States should take every action available to it to cushion the adverse consequences of the next oil disruption. A large Strategic Petroleum Reserve together with a carefully planned, well-organized allocations program worked out in cooperation with the petroleum industry offers the best short-term protection. Three benefits will result from these recommendations. First, a 2-billion-barrel reserve of petroleum stocks, in combination with a domestic-production level of roughly 10 million barrels a day would provide the nation with petroleum supplies for nearly a year. With such a reserve capability the United States would have an impressive capacity to sustain stability in the face of a short-term oil disruption. Given the economic situation in many of the petroleum-exporting countries, the

United States should be easily capable of weathering an oil disruption without major economic or political instability.

Second, a supply of stocks capable of meeting the nation's petroleum needs for nearly a year with a workable program to assure that they will be used as a balance wheel to cover sudden shortages provides powerful psychological leverage against calculated disruptions by the exporting nations. Certainly it would cause rational decision makers in the petroleum-exporting countries to consider carefully before formulating strategies that counted on a short-term disruption to achieve political or economic goals. Large petroleum stocks, in combination with an allocation or rationing program, would provide the United States with powerful negotiating leverage vis-à-vis oil-exporting nations.

Third, the recommendations made here offer the nation major domestic political and economic advantages. If the past is prologue to the future, there is a nearly immutable law of American politics that underlies these short-term recommendations: sudden, severe disruptions in physical activities that are critical to the well-being of the United States produce a near-instantaneous public demand for government action. It is unlikely that any president or Congress can withstand the pressure for action that will come with the next oil disruption. Without an allocation system and a large petroleum reserve, disorder is a nearly inevitable result. If the American population is not assured of the nation's capability to meet and manage an oil-import shortage in a stable, orderly fashion, the pattern of hoarding and supply disruption that the nation has twice experienced will almost certainly be repeated, with damaging effects to the nation's economy, the political system, and the people's confidence. If it is correct to assume that the first response of the government would be to seek to implement some kind of allocation or rationing program, then prior planning and a well-publicized program are compelling needs. In the absence of such a program, the disorder attendant on a crash rationing program would simply feed public insecurity and erode confidence both in government and in the energy industry.

The Reagan administration has indicated that its "plan" is to allow the marketplace to handle a short-term oil disruption. The political viability of such an approach can be assessed by speculating on what the public response would be if the president were to say: "The government isn't doing anything. We're leaving it to Exxon and the other oil companies."

LIQUID-FUEL OPTIONS

Recommendation 3: *The government should act on an urgent basis to subsidize an alternative domestic liquid-fuel production capability of 250,000 to 500,000 barrels per day.*

Recommendation 4: *The government-subsidized program of alternative liquid-fuels production should be carried out by the large energy companies.*

In the mid-term future the nation's transportation system will remain dependent on liquid fuels. That future will be more secure if the United States has available the demonstrated capacity to produce liquid fuels from nonoil sources on a commercial scale. At present alternative-liquids technologies are in the potential subsector. There is no agreement either on how long it will take to develop a commercial production capability or on what the fuels will cost.

An accelerated commercial demonstration program for producing alternative liquid fuels would provide a number of benefits. First are the potential economic benefits if each barrel of alternative liquid fuels can be produced at or below the full cost of an imported barrel of oil, which has been characterized by James K. Harlan as "the market price plus the additional cost reflected by an oil import premium" (Harlan, 1982, p. 4).

Estimates of the oil-import premium range from $4 to $40 per barrel as follows:

The concept of an oil import premium is becoming widely accepted. The elements of an oil import premium vary depending upon the assessment but generally include: 1) an oil pricing benefit whereby lower demand for oil imports reduces pressures on world market supplies and leads to lower prices, 2) domestic economic

300

benefits include improved balance of payments arising from reduced dollar outflows for oil purchases, reduced inflation when price increases for petroleum produces occur, and expanded domestic employment, 3) energy security benefits arise from the reduced role of unreliable oil supplies in domestic and free world economies. (Harlan, pp. 19–20)

Second, important information benefits are to be gained from the development of an alternative liquid-fuel program. Information about liquid-fuel options will reduce the uncertainties faced by private-sector investors in the future in assessing alternative investment choices. Improved information will also help the energy policy system rank the liquid-fuel options and clarify the choices available to the nation. Information gained from the first few commercial plants will not only reduce uncertainties about the costs of production but also provide essential information on the environmental and socioeconomic impacts of such plants (Harlan, pp. 28–34). Such information is critical to developing the consensus necessary should the nation be required to move to large-scale commercial production in the future.

The third benefit of a demonstration program has been termed by Harlan as the "surge deployment benefits." Surge deployment benefits are analogous to insurance payoffs from crisis events such as oil embargoes (Harlan, pp. 34–64). Those benefits would come in experience acquired by energy companies and architectural and engineering firms in building the first generation of alternative liquid-fuel plants. This initial construction phase puts the firms involved in a position to respond more efficiently and rapidly to a future need for broad-scale commercial development of alternative liquid-fuel plants.

It must be emphasized that a demonstration program offers no possibility of substantial new sources of liquids in the short term. A period of five to seven years will be necessary to construct the first generation commercial-scale facilities and begin production. Once production is proven, a similar period of time would likely be required to begin the first phase of an alternative-liquids industry.

Two potentially large sources of alternative liquid fuels stand out: synfuels from coal and oil shale and liquids from natural gas. A number of coal and oil-shale synthetic liquids technologies have been tested on a small scale. At the present time all the technologies are in the potential subsector; that is, there is disagreement in the scientific-technical community about when they can be made available and what their costs will be.

The main uncertainty about these technologies is the nature of the problems associated with scaling up synthetic liquids plants. No science of "scaling up" exists. The only way to determine the effectiveness of new technologies, their costs, and their environmental impacts is to build the plants and test them on a commercial scale.

Commercial-scale synfuels plants are very expensive. The Office of Technology Assessment estimates the cost of a commercial-scale plant at between $3 and $5 billion, with a high probability of cost overruns (U.S. Office of Technology Assessment, p. 173). The construction of commercial plants requires not only long lead times but also large-scale technical, economic, and managerial capabilities. With such long lead times and high initial costs, even the largest energy corporations have found it difficult to justify the risks; witness the shutdown of Exxon's Colony Oil Shale Project.

Corporate managers are understandably concerned about the technical and economic uncertainties of scaling up such plants:

> With new and unproven processes, there is always the risk that first generation commercial-scale plants will perform poorly and never reach their rated output capacity. Because return on capital investment is such a large part of the cost of synfuels plants, poor performance in any part of the plant, including emissions and environmental controls, has a significant impact on synfuels costs. For example, synfuels coming from a plant that operates at 50 percent of its rated capacity will cost 60 to 70 percent more than if the plant is operated at 90 percent of capacity. (U.S. Office of Technology Assessment, p. 174)

Similarly a drop in the world price of oil such as occurred in 1982 and 1983 can suddenly make the economics of synfuels plants very unattractive. A synfuels process that is profitable at a given price can become uneconomic with a small drop in oil prices. Similarly, synfuel producers must be concerned about the possibility of a calculated price reduction by oil exporters. Much Middle East oil has a production cost of about one dollar a barrel. If exporters were to see synfuels as a threat to sales, they could make the alternative-liquids plants uncompetitive by cutting the prices of exported oil below production costs for the alternatives. Only a national policy that commits federal revenues to subsidizing a major part of the program can ensure expedited commercial development and testing of synfuels plants.

With all these problems, there remains no alternative to building such plants, or at least commercial-scale modules for such plants, if an informed determination is to be made about their technological, economic, and environmental functioning.

The second potentially large-scale alternative liquid-fuel option is the conversion of natural gas into transportation liquids. Some resource estimators believe that huge new quantities of gas are recoverable from unconventional sources such as gas in tight sands and from pockets deep in the ground. At present there is great uncertainty about quantities and recovery costs. Should the optimists be correct, however, natural gas could become a major domestic resource for transportation liquids. At present, there are no large-scale conversion plants in the United States. The facilities that are in operation are devoted to the conversion of natural gas into methanol, primarily for chemical purposes.

The development of commercial-scale conversion plants offers international benefits. There are huge reserves of natural gas around the world that are not presently available to the energy market because there is no efficient means of transporting natural gas. Plants that could economically convert natural gas into transportation liquids could substantially expand the world supply of liquid fuels. These gas-to-liquids plants could also enable the nation to take advantage of technologies that are being developed to convert coal into synthetic gas (syngas). Gas-to-liquids conversion plants could be the final step in a coal-to-syngas and syngas-to-transportation-liquids process.

Conversion of coal, oil shale, and natural gas into transportation liquids is presently inhibited by the unpredictability of scale-up problems. A federally supported program would allow for rapid resolution of such problems.

A government program aimed at underwriting a demonstration conversion program at a scale of 250,000 to 500,000 barrels a day has potentially large national-security benefits for the nation (Harlan, 1982, pp. 164–218). It is important to emphasize, however, that such a program does not and should not imply detailed public management of the program. Quite the contrary. Government efforts to select "winner technologies" and to provide detailed management of those technologies in commercial competition have generally been unsuccessful (Nelson and Langlois, 1983). Federal underwriting of a transportation-liquids demonstration-production program should be confined to absorbing the major portion of the economic risk, leaving the management of the projects to the private-sector concerns that will build and operate the plants (Harlan, pp. 315–24). One attractive approach to such a program would be to have government simply provide a guaranteed market at a guaranteed price for the production from the plants. Under these arrangements if the price of imported oil rose above the negotiated price established for the production from alternative liquid facilities, there would be no cost to the government. If the price of imported oil should drop, the cost to the government would be the difference between the agreed-upon production price and the price of the imported oil. Such an arrangement would require a specific commitment by the Congress and the president that the federal agencies providing the subsidy would not be held responsible for a detailed accounting of the plants' construction and operating costs. When federal agencies are held responsible for detailed management, construction is inevitably slowed. With the high cost of capital any delay can be very expensive. Further, it must be emphasized that the goal here is to provide liquid-fuel options primarily for reasons of national security. Such a government-subsidized program should be viewed as an insurance program. That is, the government's cost should be viewed in the same way that the individual views insurance premiums.

The Synthetic Fuels Corporation offers a vehicle for carrying out a government subsidy program such as that described above. What is needed is a modification of the Energy Security Act to (1) expand the Synthetic Fuels Corporation's purview to include natural-gas-conversion facilities and (2) mandate the government's intention to absorb the financial risks without involving itself in management of the program.

The development of the plants should be carried out primarily by the large energy companies. Specifically, the Synthetic Fuels Corporation should supply substantial subsidies to such large companies as Exxon, Shell, and Gulf to encourage them to move rapidly in carrying out commercial tests. Large energy companies should be the primary focus of federal subsidies for the following reasons. First, all potentially available options require large facilities. Effective management of those facilities requires the technical skills and organizational-managerial competence of the very large companies and of large architectural and engineering firms. Second, the large companies can utilize the experience gained from the initial demonstration programs to move rapidly in the event of a need for a major national effort to develop a large-scale synthetic or conversion industry.

This subsidy proposal will no doubt offend both conservatives and liberals. Conservatives will view it as unwarranted government involvement in the private sector, and liberals will find subsidies for large corporations distasteful. The course recommended here rests on the conviction that the large companies offer the best means of providing the nation with a capability of moving rapidly to a large commercial production system should that become necessary at some time in the future.

Finally, one of the critical ingredients in such a demonstration program is a carefully monitored and assessed environmental program, carried out by independent third-party organizations. A broad base of information about the environmental impacts of alternative liquid fuels, collected in a manner acceptable to actors and interests concerned with environmental values, is necessary to building the consensus that will allow the nation to move rapidly to a broad-based alternative liquids-production capability.

LEGISLATIVE-MANAGEMENT ACTIONS

This book has emphasized the importance of maintaining a stable environment for energy decision making and management. There are compelling advantages in using public policy to provide a stable environment in which the energy marketplace can operate. The president and Congress must reaffirm, expand, and in some instances modify the boundaries presently established in legislation. The following recommendations will help assure the establishment of this stable environment.

Recommendation 5: *The present program of tax credits for residential and commercial energy conservation activities should be extended to 1995.*

The potential energy savings that can come from retrofitting existing buildings has a large and continuing contribution to make to energy conservation. Residential and commercial energy conservation suffer from serious market imperfections, inadequate information, and problems of access to funding. Extending conservation tax credits will provide an assured future for those businesses that provide conservation services and products. Public policy should assure a relatively long-term, stable market for them. The energy benefits of improved residential and commercial conservation, when set off against the relatively low cost to the Treasury, makes a compelling case for tax credits. Indeed, it would be beneficial to increase tax credits to allow 50 percent of retrofit costs to be written off.

Recommendation 6: *Present solar tax credits should be extended to 1995. The ceiling expenditure for solar technologies against which the tax credits can be charged should be increased from the present $10,000 to $30,000.*

The focus of solar tax credits should be placed on individual households. Since the Congress passed the initial program of solar tax credits, two circumstances have changed. First, inflation has increased the cost of solar installations. Second, solar technologies have been developed and improved. For example, initially the wind generators available to homeowners were not large enough to meet the total electricity needs of the average household. A new generation of wind generators capable of meeting those needs is coming on line, at a cost of $25,000. Solar tax credits should reflect these changes.

Most solar technologies are being produced and marketed by relatively small companies. Tax credits are critical to the maintenance of a market for these producers. To stimulate investment in the production of wind generators and other small-scale solar energy systems, investors need the assurance of tax credits until at least 1995.

Recommendation 7: *Federal financial support for state environmental, reclamation, and energy conservation agencies should be returned to the level of 1980.*

The state agencies that have primary responsibility for enforcing environmental and reclamation regulations are critical to ensuring that new energy activities will be carried out in an environmentally acceptable manner. The record is clear that the states are either unwilling or unable to underwrite enforcement costs, yet environmental protection remains a widely accepted goal in the United States. Federal funding, therefore, should assure that adequate environmental-protection activities will be pursued in the states.

Similarly, state conservation agencies have carried out two important functions. First, they have played a major role in communicating the advantages of energy conservation to the residents of their states. Second, they have been the vehicles whereby low-income families have received the financial support necessary to carry out energy conservation activities in their homes. The advantages of energy conservation are so great that the state conservation agencies should be once again funded at a level equal in constant dollars to that of 1980.

Recommendation 8: *The strength and viability of the Department of Energy and the Environmental Protection Agency must be established or reestablished.*

Stable energy management over the long term requires a viable, professionally competent Department of Energy. Energy is so important to the nation that it must be represented by a cabinet-level department. For its part the DOE must achieve high-quality, professional competence in three areas. First, it must develop the professional staff necessary to manage the R&D program recommended below. Second, it must continue to improve its energy information collection. Stable management of energy requires disinterested information provided on a consistent and reliable basis. Third, the DOE must play the cen-

307

tral role in the federal government for anticipating future energy needs and defining future energy policy options. Thus the department needs to build a strong energy policy-planning capability.

The primary need in the EPA is a rebuilding of its enforcement capability. With environmental protection a continuing major national goal, it is imperative that the agency have a competent enforcement capability to ensure achievement of that goal. The EPA also needs to rebuild its capability to carry out and sustain an environmental R&D program. Competent enforcement of environmental regulations must have a strong scientific-technical base. That base can be provided by a substantial energy-oriented environmental R&D program managed by the agency.

Recommendation 9: *The Department of the Interior should establish a single energy and minerals management bureau to manage the leasing and production of energy on public lands. That agency should pursue an "elephant hunting" strategy in its oil-and-gas leasing program on the outer continental shelf.*

Oil-and-gas experts assume that more than 50 percent of the oil and gas yet to be discovered in the United States lie under public lands and on the outer continental shelf. The fragmentation of energy management within the Department of the Interior has for years been a major impediment to the effective pursuit and competent management of energy resources on public lands. A single bureau responsible for energy management within the DOI is critical to the orderly and stable management of federal energy resources.

The leasing program on the OCS proposed by the Reagan administration, which allows the private sector to select and lease tracts that it believes have the greatest potential, should be pursued vigorously. The critical ingredient for making this process work is the confidence of the states and environmental and fishing interests that oil-and-gas exploration and production will be carried out in a prudent and environmentally safe manner. A single professionally competent energy-management organization that has environmental protection as a primary concern is critical to the accelerated development of both onshore and OCS federal lands.

Recommendation 10: *The Congress should promote the development of a permanent storage system for nuclear wastes.*

The nation's inability to develop a strategy for handling nuclear wastes is unacceptable. Whether or not the nation experiences a new wave of nuclear power generation in the future, it faces a growing problem in the nuclear wastes generated by existing nuclear power plants. Clearly, there are no easy choices, but the Congress must act to lay out a more detailed program for nuclear waste management.

Recommendation 11: *The Congress should extend the provision in the Energy Policy and Conservation Act of 1975 to require continuous improvement in the fuel efficiency of automobiles. The legislation should require that the fleet average performance of cars reach at least 35 miles per gallon by 1990.*

Clearly the most important incentive for increased efficiency in American cars over the last decade has been the increased price of fuel. With the downturn in petroleum prices in 1982 and 1983, however, the incentive for increased automobile efficiency eroded. The nation's critical energy problem over the midterm is its dependence on imported petroleum. The advantages of increased automobile efficiency are so compelling that legislative mandate is necessary to assure continued improvements in efficiency. Such a requirement would have the additional advantage of providing a stable market for the auto industry.

ENERGY RESEARCH AND DEVELOPMENT ACTIONS

The following recommendations for federal energy R&D are based on the assumption that there is not sufficient private-sector incentive to pursue effective R&D.

Recommendation 12: *The federal government should carry out a sustained program of research and development on solar photovoltaic technologies.*

Low-cost solar photovoltaic electricity generation has so many potential advantages that it begs for sustained, continuing research. Of the various solar options it is the only one that can be characterized as a relatively high technology development. The other solar options appear either to be undergoing adequate development in the private sector, as in wind power, or to offer such potentially limited benefits that federal support seems unwarranted, as in the utilization of ocean thermal gradients.

Recommendation 13: *Federal support for energy end-use conservation research and development focused on the residential-commercial sector should be returned to the level of 1980.*

Residential-commercial conservation technologies offer significant opportunities for energy saving. The fragmented character of the industry that both supplies and builds or upgrades residential commercial buildings makes it unlikely that the private sector will provide sustained R&D focused on more efficient technology. In the absence of a major private-sector effort to upgrade residential-commercial conservation technologies, a significant continuing federal program is necessary.

Recommendation 14: *A sustained program of federal research and development on coal should be maintained.*

Work on both advanced coal synthetic technologies and the generic science upon which those technologies are based, as well as work on various new combustion technologies, holds the promise of many benefits. Efficient conversion of coal into liquid or gaseous fuel has so many long-term benefits, given the size of the nation's coal resource base, that a high-quality technical competence in this area is an absolute necessity. Similarly, new combustion technologies would offer the possibility of providing many ways of making coal-produced energy available to the nation while mitigating many environmental costs. For example, the development of efficient low-cost fluidized beds would be of major environmental benefit to the nation.

Recommendation 15: *Federal support for research and development in conventional electricity generation technologies should be terminated.*

The Electric Power Research Institute and the various large equipment suppliers should provide the necessary research and development capability of traditional steam electric power plants to meet the nation's future needs.

Recommendation 16: *The Clinch River breeder-reactor program should be terminated, and a substantial program of research and development in fast breeders should be substituted.*

Given the nation's uncertain energy future, it is not prudent to eliminate the focus on the breeder reactor. Its potential for providing a major source of electrical power is great. The controversy over the Clinch River reactor, however, has polarized its defenders and opponents, and progress has virtually halted. Further, it has diverted attention from a broader focus on the nation's overall energy needs—for several years of the Carter administration it was impossible to pass an energy R&D budget because of the controversy over the Clinch River reactor. It seems clear that if there is to be a future for breeder reactors new directions must be found.

Many critics of the Clinch River reactor argue that it is outmoded. We do not seek to make any judgment on that view, but it is clear that such criticism has been a factor in the acrimonious debate. The appropriate route for the nation would be to initiate a broader-based R&D program focused on breeder reactors. Such a program should keep alive the technical capability in terms of people and organizations necessary to allow the nation to move to breeders in the event of a decision to do so in the future.

Recommendation 17: *The present fusion research and development program should be sustained.*

Proponents of fusion energy argue that it has many of the long-term advantages of the breeder reactor without some of the latter's potentially negative impacts. The fusion program is at such an early stage of development that no determination can be made among the claims and counterclaims. The fusion potential, however, should be pursued until such a determination can be made.

Recommendation 18: *The EPA's energy-related program in environmental research and development should be re-established at a constant-dollar level equal to that of 1980.*

Activities in almost all of the nation's available and potential sources of energy have environmental consequences. No prudent society can afford to follow a strategy that does not seek to expand the understanding of those consequences. That understanding provides a basis for informed regulation as well as means whereby negative consequences can be mitigated.

Recommendation 19: *The Department of Energy should carry on a small-scale funding program for energy policy research outside the federal government.*

There is a continuing need for the nation to understand its energy system, anticipate future developments, and inform the American population with regard to that system. A relatively low-cost program of energy policy research with components in various public and private sectors would make an important contribution to ensuring stable management of energy over the long term. Energy policy research carried out by non-federal, nonenergy industry sources provides an independent check on those organizations.

Recommendation 20: *The present DOE program of support for long-range research should be continued. One focus of this program should be research on long-range energy demand.*

By general agreement, long-range research is an appropriate government responsibility because the payoffs are too uncertain and too distant for the private sector to fund. The goal of research in long-range demand should be to improve the efficient matching of fuel types with end uses.

CONCLUSIONS

The recommendations in this chapter share a central theme: a stable energy future for the United States requires a cooperative relationship among participants in government, the private sector, and the nonprofit research institutions. During the 1970s and 1980s the nation moved from a stable environment for the management of fuels to detailed intervention by the government in management of the energy system to the present hands-off policy. The role of public energy policy should be to provide the private sector with a stable environment in which to work. In other words, the role of public policy should be to provide a set of dependable, clearly defined conditions under which the marketplace is allowed to operate. The use of public policy for that purpose is the traditional pattern in the United States. A secure energy future makes it imperative that the nation return to that pattern as quickly as possible.

References

CHAPTER 1

Barnett, Richard J. 1980. *The Lean Years: Politics in an Age of Scarcity*. New York: Simon and Schuster.

Carson, Rachael. 1962. *Silent Spring*. Boston: Houghton Mifflin.

Chapman, Duane. 1974. "Electricity in the United States." In Edward W. Erickson and Leonard Wavermen, eds. *The Energy Question: An International Failure of Policy*, pp. 77–96. Toronto: University of Toronto Press.

Crabbe, David, and Richard McBride, eds. 1979. *The World Energy Book*. Cambridge, Mass.: MIT Press.

Davis, David H. 1982. *Energy Politics*. 3d ed. New York: St. Martin's Press.

Dorf, Richard C. 1978. *Energy, Resources, and Policy*. Reading, Mass.: Addison-Wesley.

Edgmon, Terry D. 1979. "Organizing for Energy Policy and Administration." In Robert Lawrence, ed. *New Dimensions to Energy Policy*, pp. 81–92. Lexington, Mass.: D. C. Heath.

Ford Foundation Energy Policy Project. 1974a. *Exploring Energy Choices: A Preliminary Report*. Washington, D.C.: Ford Foundation.

———. 1974b. *A Time to Choose: America's Energy Future*. Cambridge, Mass.: Ballinger.

Friends of the Earth. 1982. *Ronald Reagan and the American Environment*. San Francisco: Friends of the Earth.

Garvey, Gerald. 1975. "Environmentalism Versus Energy Development: The Constitutional Background to Environmental Administration." *Public Administration Review* 35 (July–August):328–33.

Jones, Charles O. 1977. *An Introduction to the Study of Public Policy*. 2d ed. North Scituate, Mass.: Duxbury Press.

Kash, Don E., et al. 1976. *Our Energy Future*. Norman: University of Oklahoma Press.

Mancke, Richard B. 1974. *The Failure of U.S. Energy Policy*. New York: Columbia University Press.

National Petroleum Council. 1971. *U.S. Energy Outlook: An Initial Appraisal, 1971–1975*. Washington, D.C.: National Petroleum Council.

Palmer, John L., and Isabel V. Sawhill, eds. 1982. *The Reagan Experi-*

ment. Washington, D.C.: Urban Institute Press.

Roberts, Marc. 1973. "Is There an Energy Crisis?" *Public Interest* 31 (Spring):17–37.

Ross, Marc H., and Robert H. Williams. 1981. *Our Energy: Regaining Control.* New York: McGraw-Hill.

Schurr, Sam H., et al. 1979. *Energy in America's Future.* Baltimore, Md.: John Hopkins University Press.

U.S. Congress, House, Committee on Science and Astronautics, Subcommittee on Energy. 1974. *Energy Policy and Resource Management.* Washington, D.C.: U.S. Government Printing Office.

U.S. Federal Energy Administration. 1974. *Project Independence Blueprint.* Washington, D.C.: U.S. Government Printing Office.

Wilbanks, Thomas J. 1981. *Building a Consensus About Energy Technologies.* Oak Ridge, Tenn.: Oak Ridge National Laboratory.

Wolozin, Harold, ed. 1974. *Energy and the Environment.* Morristown, N.J.: General Learning Press.

CHAPTER 2

Boorstin, Daniel J. 1978. *The Republic of Technology.* New York: Harper and Row.

Braybrooke, David, and Charles E. Lindblom. 1970. *A Strategy of Decision.* New York: Free Press.

Coates, Joseph F. 1979. "What Is a Public Policy Issue? An Advisory Essay." *Interdisciplinary Science Review* 4 (March):27–44.

Easton, David. 1965. *A Framework for Political Analysis.* Englewood Cliffs, N.J.: Prentice-Hall.

Gilinsky, Victor. 1980. "The Impact of Three Mile Island." *Bulletin of the Atomic Scientists* 36 (January):18–20.

Kash, Don E., et al. 1973. *Energy Under the Oceans.* Norman: University of Oklahoma Press.

Lamm, Richard. 1976. "States Rights vs. National Energy Needs." *Natural Resources Journal* 9:41–48.

Landsberg, Hans H. 1980. "Let's All Play Energy Policy!" *Daedalus* 109 (Summer):71–84.

Lindberg, Leon N. 1977. "Energy Policy and the Politics of Economic Development." *Comparative Political Studies* 10 (October):355–82.

McFarland, Andrew S. 1976. *Public Interest Lobbies: Decision Making on Energy.* Washington, D.C.: American Enterprise Institute for Public Policy Research.

Mancke, Richard B. 1974. *The Failure of U.S. Energy Policy.* New York: Columbia University Press.

Mazur, Allan. 1981. *The Dynamics of Technical Controversy.* Washington, D.C.: Communications Press.

Nelkin, Dorothy. 1981. "Science, Technology, and Political Conflict: Analyzing the Issues." In Dorothy Nelkin, ed. *Politics of Technical Decisions,* pp. 9–24. Beverly Hills, Calif: Sage.

Noble, David F. 1977. *America by Design.* New York: Alfred A. Knopf.

Nowotny, Helga. 1981. "Experts and Their Expertise: On the Changing Relationship Between Experts and Their Public." *Bulletin of Science, Technology & Society* 1, no. 3:235–41.

Price, Don K. 1954. *Government and Science: Their Dynamic Relations in American Democracy.* New York: New York University Press.

Rycroft, Robert W., and Joseph S. Szyliowicz. 1980. "The Technological Dimension of Decision-Making: The Case of the Aswan High Dam." *World Politics* 33 (October):36–61.

Wilson, James Q. 1975. "The Rise of the Bureaucratic State." *Public Interest* 41 (Fall):77–103.

CHAPTER 3

Bupp, I. C. 1979. "The Nuclear Stalemate." In Robert Stobaugh and Daniel Yergin, eds. *Energy Future,* pp. 108–34. New York: Random House.

——, and Frank Schuller. 1979. "Natural Gas: How to Slice a Shrinking Pie." In Robert Stobaugh and Daniel Yergin, eds. *Energy Future,* pp. 56–78. New York: Random House.

Congressional Quarterly, Inc. 1981. *Energy Policy.* 2d ed. Washington, D.C.: Congressional Quarterly, Inc.

——. 1982. *Energy Issues: New Directions and Goals.* Washington, D.C.: Congressional Quarterly, Inc.

Darmstadter, Joel, Joy Dunkerly, and Jack Alterman. 1977. *How Industrial Societies Use Energy.* Baltimore, Md.: Johns Hopkins University Press.

Davis, David H. 1974. *Energy Politics.* New York: St. Martin's Press. 2d ed., 1978. 3d ed., 1982.

Executive Office of the President. 1973. "Energy Policy: The President's Message to the Congress." *Weekly Compilation of Presidential Documents* 9, no. 16 (April 23):389.

——. 1977. *The National Energy Plan.* Washington, D.C.: U.S. Government Printing Office.

Ford Foundation Energy Policy Project. 1974a. *Exploring Energy Choices: A Preliminary Report.* Washington, D.C.: Ford Foundation.

——. 1974b. *A Time to Choose: America's Energy Future.* Cambridge, Mass.: Ballinger.

Freehling, Robert J. 1973. "Health and Safety Regulations." *American*

Law of Mining 3:713–36.

Hoerr, John. 1974. "Coal and the Mine Workers." *Atlantic Monthly* 235 (March):10–23.

Kash, Don E., et al. 1973. *Energy Under the Oceans.* Norman: University of Oklahoma Press.

Krenz, Jerrold H. 1980. *Energy: From Opulence to Sufficiency.* New York: Praeger.

Lapp, Ralph E. 1976. *America's Energy.* Greenwich, Conn.: Reddy Communications.

Lave, Lester B., and Gilbert S. Omenn. 1981. *Clearing the Air: Reforming the Clean Air Act.* Washington, D.C.: Brookings Institution.

National Research Council. 1979. *Energy in Transition. 1985–2010.* San Francisco: W. H. Freeman.

Nuclear Energy Policy Study Group. 1977. *Nuclear Power: Issues and Choices.* Cambridge, Mass.: Ballinger.

O'Toole, James. 1976. *Energy and Social Change.* Cambridge, Mass.: MIT Press.

Rosenbaum, Walter A. 1977. *The Politics of Environmental Concern.* 2d ed. New York: Praeger.

Stobaugh, Robert. 1979. "After the Peak: The Threat of Imported Oil." In Robert Stobaugh and Daniel Yergin, eds. *Energy Future,* pp. 16–55. New York: Random House.

U.S. Department of the Interior. 1980. *The Use of Best Available and Safest Technologies (BAST) During Oil and Gas Drilling and Producing Operations on the Outer Continental Shelf.* Washington, D.C.: U.S. Department of the Interior, U.S. Geological Survey.

U.S. Federal Power Commission. 1970. *The 1970 National Power Survey Guidelines for Growth of the Electric Power Industry, Part 2.* Washington, D.C.: U.S. Government Printing Office.

U.S. General Accounting Office. 1977. *U.S. Coal Development, Promises, Uncertainties: Report to the Congress.* Washington, D.C.: U.S. General Accounting Office.

CHAPTER 4

Bereano, Philip L. 1976. *Technology as a Social and Political Phenomenon.* New York: Wiley.

Congressional Quarterly, Inc. 1982. *Environmental Issues: Prospects and Problems.* Washington, D.C.: Congressional Quarterly, Inc.

De Marchi, Neil. 1981. "The Ford Administration: Energy as a Political Good." In Craufurd D. Goodwin, ed. *Energy Policy in Perspective: Today's Problems, Yesterday's Solutions,* pp. 475–545. Washington, D.C.: Brookings Institution.

Garvey, Gerald. 1975. "Environmentalism Versus Energy Development: The Constitutional Background to Environmental Administration." *Public Administration Review* 35 (July–August):328–33.

Grafton, Carl. 1975. "The Creation of Federal Agencies," *Administration and Society* 7 (November):328–65.

Hahn, Walter, 1975. "Foreword." In Sherry R. Arnstein and Alexander N. Christakis, eds. *Perspectives on Technology Assessment*, pp. ix–xv. Jerusalem: Science and Technology Publishers.

Institute for Energy Analysis. 1979. *Economic and Environmental Impacts of a Nuclear Moratorium, 1985–2010.* Cambridge, Mass.: MIT Press.

Koreisha, Sergio, and Robert Stobaugh. 1979. "Appendix I: Limits to Models." In Robert Stobaugh and Daniel Yergin, eds. *Energy Future*, pp. 235–365. New York: Random House.

Landsberg, Hans H. 1980. "Let's All Play Energy Policy!" *Daedalus* 109 (Summer):71–84.

———, and Joseph M. Dukert. 1981. *High Energy Costs: Uneven, Unfair, Unavoidable?* Baltimore, Md.: Johns Hopkins University Press.

Lindberg, Leon N. 1977. "Energy Policy and the Politics of Economic Development." *Comparative Political Studies* 10 (October):355–82.

Lovins, Amory B. 1977. *Soft Energy Paths: Toward a Durable Peace.* Cambridge, Mass.: Ballinger.

Meadows, Donella H., et al. 1972. *The Limits to Growth.* New York: New American Library.

CHAPTERS 5 TO 7

Atomic Energy Commission. 1973. *The Nation's Energy Future: A Report to Richard M. Nixon, President of the United States.* Washington, D.C.: U.S. Government Printing Office.

———. 1974. *Draft Environmental Statement: Liquid Metal Fast Breeder Reactor Program.* Vol. 1, *Alternative Technology Options.* Washington, D.C.: U.S. Government Printing Office.

Atomic Industrial Forum, Inc. 1974. "Comparison of Fuels Used in Power Plants." *Background INFO.* New York: Atomic Industrial Forum, Inc.

Beck, Robert J. 1983. "Demand, Imports to Rise in 83; Production to Slip." *Oil and Gas Journal*, January 31, p. 76.

Bodle, W. W., and B. E. Eakin. 1971. "Predicting Properties for LNG Operations." *Proceedings of the 50th Annual Convention, Technical Papers.* National Gas Producers Association, Houston, Texas, March 17–19.

Bureau of National Affairs. 1982a. *Energy Users Report* 11 (March

11):242.

——. 1982b. *Energy Users Report* 36 (September 9).

California Resources Agency. 1974. *California State Water Project: Annual Report, 1973.* Sacramento: California Resources Agency.

Creager, William P., and Joel D. Justin. 1950. *Hydroelectric Handbook.* New York: Wiley.

Garrett Research and Development Company, Inc. N.d. *Solid Waste Disposal and Resource Recovery.* La Verne, Calif.: Garrett Research and Development Company, Inc.

General Electric Company, Power Systems Sector. 1981. *United States Energy Data Book.* Fairfield, Conn.: General Electric.

Gillette, Robert. 1974. "Oil and Gas Resources: Did USGS Gush Too High?" *Science* 180 (July 12):127–30.

Gouse, S. William, Jr., and Edward S. Rubin. 1973. *A Program of Research, Development, and Demonstration for Enhancing Coal Utilization to Meet National Emergency Needs.* Results of the Carnegie-Mellon University Workshop on Advanced Coal Technology. N.p.

INFO. 1974. *Public Affairs and Information Program.* No. 70 (May). New York: Atomic Industrial Forum, Inc.

Kash, Don E., et al. 1976. *Our Energy Future.* Norman: University of Oklahoma Press.

National Petroleum Council. 1972. *U.S. Energy Outlook: An Initial Appraisal.* Washington, D.C.: National Petroleum Council.

Science and Public Policy Program, University of Oklahoma. 1975. *Energy Alternatives.* Washington, D.C.: U.S. Government Printing Office.

Surface Mining Control and Reclamation Act. 1977. 91 Stat. 445.

U.S. Bureau of Mines. 1950. *Potential Oil Recovery by Waterflooding Reservoirs Being Produced by Primary Measures.* Information Circular 8455. Washington, D.C.: U.S. Government Printing Office.

U.S. Congress, Office of Technology Assessment. 1981. *Patterns and Trends in Federal Coal Lease Ownership, 1950–1980: A Technical Memorandum.* Washington, D.C.: U.S. Government Printing Office.

U.S. Congressional Research Service. 1980. *The Energy Fact Book.* Prepared at the request of the Subcommittee on Energy and Power of the Committee on Interstate and Foreign Commerce. Committee Print 96-IFC-60. 96th Cong., 2d sess. Washington, D.C.: U.S. Government Printing Office.

U.S. Department of Commerce, Bureau of the Census. 1979. *Statistical Abstracts of the United States, 1979.* 100th ed. Washington, D.C.: U.S. Government Printing Office.

——. 1981. *Statistical Abstracts of the United States, 1981.* 102d ed.

Washington, D.C.: U.S. Government Printing Office.

U.S. Bureau of Land Management. 1974. *Draft Environmental Impact Statement: Proposed Federal Coal Leasing Program.* 2 vols. Washington, D.C.: U.S. Government Printing Office.

U.S. Department of Energy. 1981. *U.S. Crude Oil, Natural Gas, and Natural Gas Liquid Reserves: 1979 Annual Report.* Washington, D.C.: U.S. Government Printing Office.

U.S. Department of the Interior. 1973. *Final Environmental Statement for the Prototype Oil Shale Leasing Program.* 6 vols. Washington, D.C.: U.S. Government Printing Office.

————, U.S. Geological Survey. 1981. *Estimates of Undiscovered Recoverable Conventional Resources of Oil and Gas in the United States.* Circular 860. Washington, D.C.: U.S. Government Printing Office.

Wild and Scenic Rivers Act. 1968. 82 Stat. 906.

Wildavsky, Aaron, and Ellen Tenenbaum. 1981. *The Politics of Mistrust.* Beverly Hills, Calif.: Sage.

CHAPTERS 8 TO 10

Alaska Lands Act. 1980. 94 Stat. 2371.

Alaska Natural Gas Transportation Act. 1976. 90 Stat. 2903.

Alaska Natural Gas Transportation System Approval. 1977. 91 Stat. 1268.

Andrus, Secretary of the Interior v. *Virginia Surface Mining and Reclamation Association et al.* 1980. 445 U.S. 922.

Appropriations—Atomic Energy Commission. 1970. 84 Stat. 299.

Atomic Energy Act. 1974. 89 Stat. 1111.

Beck, Robert J. 1981. "Demand, Production, Imports, All to Fall." *Oil and Gas Journal.* January 26:127.

Bupp, I. C. 1979. "The Nuclear Stalemate." In Robert Stobaugh and Daniel Yergin, eds. *Energy Future,* pp. 108–35. New York: Random House.

Bureau of National Affairs, Inc. 1980a. *Energy Users Report* 338 (January 31):11.

————. 1980b. *Energy Users Report* 340 (February 14):9.

————. 1980c. *Energy Users Report* 414 (July 16):1107.

————. 1980d. *Energy Users Report* 365 (August 7):8.

————. 1980e. *Energy Users Report.* Current Report no. 371 (September 18):17.

————. 1982. *Energy Users Report* 31 (August 5):819.

Clean Air Amendments. 1970. 84 Stat. 1676.

Clean Air Act Amendments. 1977. 91 Stat. 685.

Combined Hydrocarbon Leasing Act of 1981. 95 Stat. 1069.

Congressional Quarterly, Inc. 1975. *Congressional Quarterly Almanac, 1974.* Vol. 30. Washington, D.C.: Congressional Quarterly, Inc.

――――. 1976a. *Congressional Quarterly Weekly Report.* No. 48 (November 27):3239.

――――. 1976b. *Congressional Quarterly Almanac.* Vol. 32. Washington, D.C.: Congressional Quarterly, Inc.

――――. 1981. *Energy Policy.* 2d ed. Washington, D.C.: Congressional Quarterly, Inc.

――――. 1982. *Energy Issues: New Direction and Goals.* Washington, D.C.: Congressional Quarterly, Inc.

Congressional Quarterly Service. 1973. *Congress and the Nation.* Vol. 3. Washington, D.C.: Congressional Quarterly, Inc.

Congressional Record. 1972. P. 35031.

――――. 1973. Pp. 576–94.

Congressional Research Service. 1980. *The Energy Fact Book. Prepared at the Request of the Subcommittee on Energy and Power of the Committee on Interstate and Foreign Commerce.* Committee print 96-IFC-60. 96th Cong., 2d sess. Washington, D.C.: U.S. Government Printing Office.

Corrigan, Richard E. 1981. "Synfuels Subsidies—Reports of Their Death Were Greatly Exaggerated." *National Journal* 13 (March 14): 430.

Crude Oil Windfall Profits Tax Act. 1980. 94 Stat. 229.

Department of Energy Organization Act. 1977. 91 Stat. 565.

Emergency Daylight Saving Time Energy Conservation Act. 1973. 87 Stat. 707.

Emergency Energy Conservation Act. 1979. 93 Stat. 749.

Emergency Highway Energy Conservation Act. 1973. 87 Stat. 1046.

Emergency Natural Gas Act. 1977. 91 Stat. 4.

Emergency Petroleum Allocation Act. 1973. 87 Stat. 627.

Energy Conservation and Production Act. 1976. 90 Stat. 1125.

Energy Information Center, Inc. 1976. *The Energy Index, 1976.* New York: Environment Information Center, Inc.

Energy Policy and Conservation Act. 1975. 89 Stat. 781.

Energy Reorganization Act. 1974. 88 Stat. 1233.

Energy Security Act. 1980. 94 Stat. 611.

Energy Supply and Environmental Coordination Act. 1974. 88 Stat. 246.

Energy Tax Act. 1978. 92 Stat. 3174.

Executive Office of the President, Office of Management and Budget. 1975. *Budget of the United States Government FY1975.* Washington, D.C.: U.S. Government Printing Office.

————. 1976. *Special Analysis: Budget of the United States Government FY1976.* Washington, D.C.: U.S. Government Printing Office.

————. 1982. *Budget of the United States Government FY1982.* Washington, D.C.: U.S. Government Printing Office.

Executive Order 11703. February 8, 1973.

Executive Order 11712. April 23, 1973.

McCaslin, John C. 1981. "Torrid Drilling to Surge Past 70,000 Wells." *Oil and Gas Journal,* January 26, p. 145.

Murphy, Thomas A. 1975. "Bigness Defended." *New York Times,* April 28, sec. 4, p. 17.

Nanz, Robert. 1980. *Federal Land Availability: An Industry View.* Paper presented at the Society of Petroleum Engineers Annual Technological Conference and Exhibition, Dallas, Texas, September 23.

National Energy Conservation Policy Act. 1978. 92 Stat. 3206.

Natural Gas Policy Act. 1978. 92 Stat. 3350.

Naval Petroleum Reserves Production Act. 1976. 90 Stat. 303.

New York Times. 1975. "G. M. Executives Forecast Higher Priced '76 Models." May 1, p. 59.

Oil and Gas Journal. 1970a. "FPC Moves to Jack Permian Gas Prices." June 1, p. 41.

————. 1970b. "Gas Price-Hike of 5-10 Cents Would Spur Drilling." June 1, p. 52.

————. 1971a. "Experts See Grim Outlook for U.S. Energy Supplies." October 4, pp. 64–69.

————. 1971b. "Rush to Costly SNG Proves 'Cheap-Gas' Policy a Failure." November 15, p. 87.

————. 1972. "Suspension of Wellhead Price Controls Urged." January 24, p. 36.

————. 1973a. "Action, Not Words, Solution for Energy." June 25, p. 51.

————. 1973b. "Stevenson Bill Proposes Federal Oil, Gas Company." November 12, p. 90.

————. 1981. "Synfuels Action Continues Brisk Across U.S." February 9, p. 32.

————. 1981b. "U.S. Lets Last Big NPRA Exploration Contracts." April 6, p. 51.

Outer Continental Shelf Lands Act Amendments. 1978. 92 Stat. 629.

Pelham, Ann. 1980. "Final Vote on Windfall Profits Provisions Is Expected Soon in Senate and House." *Congressional Quarterly.* March 8, pp. 668–69.

Power Plant and Industrial Fuel Use Act. 1978. 92 Stat. 2389.

Public Utility Regulatory Policies Act. 1978. 92 Stat. 3117.

Rosenbaum, Walter A. 1981. *Energy, Politics, and Public Policy.* Washington, D.C.: Congressional Quarterly, Inc.

Solar Energy Research, Development and Demonstration Act. 1974. 88 Stat. 1431.

Supplemental Appropriations Act. 1978. 92 Stat. 107.

Surface Mining and Control and Reclamation Act. 1977. 91 Stat. 445.

Tax Reduction Act. 1975. 89 Stat. 16.

Trans-Alaska Pipeline Authorization Act. 1973. 87 Stat. 576.

U.S. Congress. 1979. Joint Hearing Before the Subcommittee on Nuclear Regulation of the Committee on Environment and Public Works, U.S. Senate and the Subcommittee on Energy and the Environment of the Committee on Interior and Insular Affairs, U.S. House of Representatives. *Report of the President's Commission on Three Mile Island Accident.* Serial no. 96-H34. 96th Cong., 1st sess.

————. 1970. *Committee on Government Operations Approving Reorganization Plan Number Three of 1970.* H. Rept. 91-1464, 91st Cong., 2d sess.

————. 1973. *Emergency Petroleum Allocation Act of 1973.* H. Rept. 93-513, 93d Cong., 1st sess.

————. 1975. *Message from the President of the United States: Veto of the Surface Minings Control and Reclamation Act of 1975.* H. Doc. 94-160, 94th Cong., 1st sess.

————. 1976. *Naval Petroleum Reserves Production Act of 1976.* H. Rept. 94-81, 94th Cong., 2d sess.

————., Committee on Interior and Insular Affairs. 1977a. *Alaska Natural Gas Transportation System-Approval.* H. Rept. 95-739, 95th Cong., 1st sess.

————. 1977b. *Communication from the Comptroller General of the United States.* H. Doc. 95-176, 95th Cong., 1st sess.

————. 1977c. *Hearings Before the Subcommittee on Energy and Power of the Committee on Interstate and Foreign Commerce to Consider HR6831, the National Energy Act.* Serial no. 95-22, 95th Cong., 1st sess.

U.S. Department of Commerce, Bureau of the Census. 1981. *Statistical Abstract of the United States, 1981.* Washington, D.C.: U.S. Government Printing Office.

U.S. Department of Energy, Energy Information Administration. 1982. *1981 Annual Report to Congress. Energy Statistics.* Vol. 2. Washington, D.C.: U.S. Government Printing Office.

U.S. Department of the Interior, United States Geological Survey. 1980a. *The Use of Best Available and Safest Technologies (BAST) During Oil and Gas Drilling and Producing Operations on the Outer Continental Shelf.* Washington, D.C.: U.S. Department of the Interior.

————, Office of the Secretary. 1980b. Letter to the President, April 4, 1980.

————, U.S. Geological Survey, Bureau of Land Management. 1981. *Compilation of Regulations Relating to Mineral Resource Activities on the Outer Continental Shelf.* Vols. 1, 2.

U.S. Nuclear Regulatory Commission. 1975. *Reactor Safety Study—An Assessment of Accident Risk in U.S. Commercial Nuclear Power.* Main Report. Springfield, Va.: National Technical Information Service.

U.S. President. 1971. "Executive Reorganization: Message to Congress" (March 25). *Weekly Compilation of Presidential Documents 7*, no. 15 (March 29):545, 552.

————. 1973a. "The Energy Emergency: Message to the Congress" (November 7). *Weekly Compilation of Presidential Documents 9*, no. 45 (November 10):1319.

————. 1973b. "Energy Policy: The President's Message to the Congress" (April 18). *Weekly Compilation of Presidential Documents 9*, no. 16 (April 23):389.

————. 1974. "Energy Crisis: Message to Congress" (January 23). *Weekly Compilation of Presidential Documents 10*, no. 4 (January 25):72.

————. 1975. "Energy and Economic Programs: President's Address to the Nation" (January 13). *Weekly Compilation of Presidential Documents 11*, no. 3 (January 20):40.

————. 1976. "Energy: Message to Congress" (February 16). *Weekly Compilation of Presidential Documents 12*, no. 9 (March 1):289.

————. 1977a. "Energy: Letter to Speaker of the House and the President of the Senate" (March 6). *Weekly Compilation of Presidential Documents 16*, no. 10 (March 10):448.

————. 1977b. "Energy Fact Sheet" (April 20). *Weekly Compilation of Presidential Documents 13*, no. 17 (April 22):582.

————. 1977c. "Energy: Message to Congress" (April 20). *Weekly Compilation of Presidential Documents 13*, no. 17 (April 25):571.

————. 1977d. "National Energy Program: Message to the Congress" (April 20). *Weekly Compilation of Presidential Documents 13*, no. 17 (April 22):566.

————. 1977e. "Energy Message to the Senate" (November 5). *Weekly Compilation of Presidential Documents 13*, no. 46 (November 14): 1726.

————. 1979. "Energy Policy: The President's Message to the Congress" (April 5). *Weekly Compilation of Presidential Documents 15*, no. 14 (April 9):609.

University of Oklahoma, Science and Public Policy Program. 1975. *Energy Alternatives.* Washington, D.C.: U.S. Government Printing Office.

Washington Post. 1973a. "Six Oil Companies Report Increase in Sales

Profit." January 26, sec. D, p. 13.

———. 1973b. "Texaco, Gulf, Continental, Phillips, Report Profits Gain." April 25, sec. E, p. 4.

———. 1980. "Better Mileage, Industrial Cooperation Sought as Strategies in War Against Japanese Imports." July 10, sec. B, p. 1.

———. 1981. "OPEC Benchmark Prices, 1973–1981." October 30, p. A5.

West Valley Demonstration Project Act. 1980. 94 Stat. 1347.

CHAPTER 11

Appropriations for the Department of Interior and Related Agencies for the Fiscal Year Ending September 30, 1982, and for Other Purposes. 1981. 95 Stat. 1391.

Barfield, Claud E. 1982. *Science Policy from Ford to Reagan: Change and Continuity.* Washington, D.C.: American Enterprise Institute for Public Policy Research.

Benjamin, Milton R. 1982. "White House Weighs Revival of S.C. Nuclear Fuel Plant." *Washington Post,* June 19, pp. A1, A6.

———. 1983. "President Sends Hill Plan to Decontrol Natural Gas Prices." *Washington Post,* March 1, p. A3.

Bromley, D. Allan. 1981. "Welcome." In Albert H. Teich, Gail J. Breslow, and Jill P. Weinberg, eds. *R&D and the New National Agenda,* pp. 8–10. Washington, D.C.: American Association for the Advancement of Science.

Bruce, James T. 1983. "A Plan Not to Plan Is the Plan." *Environment* 25 (March):6–9, 28–35.

Bureau of National Affairs, Inc. 1982. *Energy Users Report* 31 (August 5):805.

———. 1982a. *Energy Users Report* 414 (July 16):1007–1115.

———. 1981b. *Energy Users Report* 427 (October 15):1503.

Congressional Quarterly, Inc. 1982a. *Energy Issues: New Directions and Goals.* Washington, D.C.: Congressional Quarterly, Inc.

———. 1982b. *Environmental Issues: Prospects and Problems.* Washington, D.C.: Congressional Quarterly, Inc.

Corrigan, Richard. 1982a. "The Administration May Try Again to Kill Energy Aid for the Poor." *National Journal* 14 (January 23): 150–52.

———. 1982b. "Even Under Reagan Synfuels Corp. Ready with Big Bucks for Private Projects." *National Journal* 14 (February 6):228–32.

———. 1982c. "Creative Flair," *National Journal* 14 (November 6): 1906.

Eads, George. 1981. "Harnessing Regulation: The Evolving Role of White House Oversight." *Regulation* 5 (May–June):19–26.

Fried, Edward. 1982. "Energy Security and the Common Interest." *Brookings Review* 1 (Winter):10–13.

Friends of the Earth. 1982. *Ronald Reagan and the American Environment.* San Francisco: Friends of the Earth.

Goodwin, Craufurd D. 1981. "The Lessons of History." *Wilson Quarterly* 5 (Spring):91–97.

Inside E.P.A. 1982. "Gorsuch Tells States EPA Has Long-Term Goal of Ending All State Grants." Vol. 3 (September 24):3.

Jennrich, John H. 1982. "U.S. Synthetic Fuels Corp. Picks Up Speed." *Oil and Gas Journal,* May 18, p. 111.

Kash, Don E., et al. 1976. *Our Energy Future.* Norman: University of Oklahoma Press.

Kurtz, Howard. 1983. "EPA Removed More Than 50 Scientists on Conservative 'Hit List.'" *Washington Post,* March 2, p. A2.

Marshall, Eliot. 1981. "Quick March on Nuclear Licensing." *Science* 213 (August 28):983–84.

Mosher, Lawrence. 1982a. "Making Them Hire." *National Journal* 14 (March 20):514.

———. 1982b. "More Cuts in EPA Research Threaten Its Regulatory Goals, Critics Warn." *National Journal* 14 (April 10):635–39.

———. 1982c. "EPA, Critics Agree Agency Under Gorsuch Hasn't Changed Its Spots." *National Journal* 14 (November 13):1941–44.

———. 1983a. "Better Than Something." *National Journal* 15 (January 29):230.

———. 1983b. "Distrust of Gorsuch May Stymie EPA Attempt to Integrate Pollution Wars." *National Journal* 15 (February 12):322–24.

National Governors' Association and National Conference of State Legislatures. 1982. *The Proposed FY 1983 Federal Budget: Impact on the States.* Washington, D.C.: National Governors' Association and National Conference of State Legislatures.

Norman, Colin. 1981a. "Reagan Energy Plan Reluctantly Unveiled." *Science* 213 (July 31):520–22.

———. 1981b. "Reagan's Energy Policy and Other 'Myths.'" *Science* 213 (September 25):1481.

Oil and Gas Journal. 1981. "Senate Panel Backs Move of SPR off Budget." May 18, p. 28.

Plattner, Andy. 1982. "Energy Issues Shoved onto Back Burner." *Congressional Quarterly Weekly Report* 40 (May 29):1249–52.

Regens, James L., Robert W. Rycroft, and Gregory A. Daneke. 1982. "Deja Vu and Western Energy Development." In James L. Regens, Robert W. Rycroft, and Gregory A. Daneke, eds. *Energy and the*

Western United States: Politics and Development, pp. 183–88. New York: Praeger.

Shapley, Willis H., Albert H. Teich, and Jill P. Weinberg. 1982. *Research and Development: AAAS Report VII.* Washington, D.C.: American Association for the Advancement of Science.

U.S. Congress, House. 1982. *Further Continuing Resolution for Department of Energy and other Agencies for FY 1983.* Washington, D.C.: U.S. Government Printing Office.

U.S. Congressional Research Service. 1981. *The Unfolding of the Reagan Energy Program: The First Year.* Washington, D.C.: Congressional Research Service, Library of Congress.

U.S. Department of Interior, U.S. Geological Survey, Bureau of Land Management. 1981. *Compilation of Regulations Relating to Mineral Resource Activities on the Outer Continental Shelf.* Vols. 1–2. Washington, D.C.: U.S. Geological Survey.

U.S. Office of Management and Budget [Executive Office of the President]. 1983. "Special Analysis K: Research and Development." *The Budget for Fiscal Year 1984.* Washington, D.C.: U.S. Government Printing Office.

William, Juan. 1983. "Reagan Describes Decontrol of Gas as Consumer Aid." *Washington Post,* February 27, p. A8.

CHAPTERS 12 AND 13

Beck, Robert J. 1982. "U.S. Oil Demand to Fall Again by 4.2%; Imports Also Slide, but Production Up." *Oil and Gas Journal,* July 26.
———. 1983. "Demand, Imports to Rise in 83; Production to Slip." *Oil and Gas Journal,* January 31.

Bell, Daniel. 1981. "Preface." In *President's Commission for a National Agenda for the Eighties, Energy, Natural Resources and the Environment in the Eighties.* Washington, D.C.: U.S. Government Printing Office.

Editorial Research Reports. 1982. *Energy Issues: New Directions and Goals.* Washington, D.C.: Congressional Quarterly, Inc.

Harlan, James K. 1982. *Starting with Synfuels.* Cambridge, Mass.: Ballinger.

Jennrich, John H. 1982. "U.S. Synthetic Fuels Corp. Picks up Speed." *Oil and Gas Journal,* June 28, p. 111.

Marshall, Eliot. 1980. "Energy Forecasts: Sinking to New Lows." *Science* 208 (June 20):1353–56.

Nelson, Richard R., and Richard N. Langlois. 1983. "Innovation Policy: Lessons from American History." *Science* 219 (February 18): 817–18.

Stobaugh, Robert. 1979. "After the Peak: The Threat of Imported Oil." In Robert Stobaugh and Daniel Yergin, eds. *Energy Future.* New York: Random House.

U.S. Office of Technology Assessment. 1982. *Increased Automobile Fuel Efficiency and Synthetic Fuels: Alternatives for Reducing Oil Imports.* Washington, D.C.: U.S. Government Printing Office.

Yergin, Daniel. 1982. "America in the Strait of Stringency." In Daniel Yergin and Martin Hillenbrand, eds. *Global Insecurity.* Boston: Houghton Mifflin.

Index